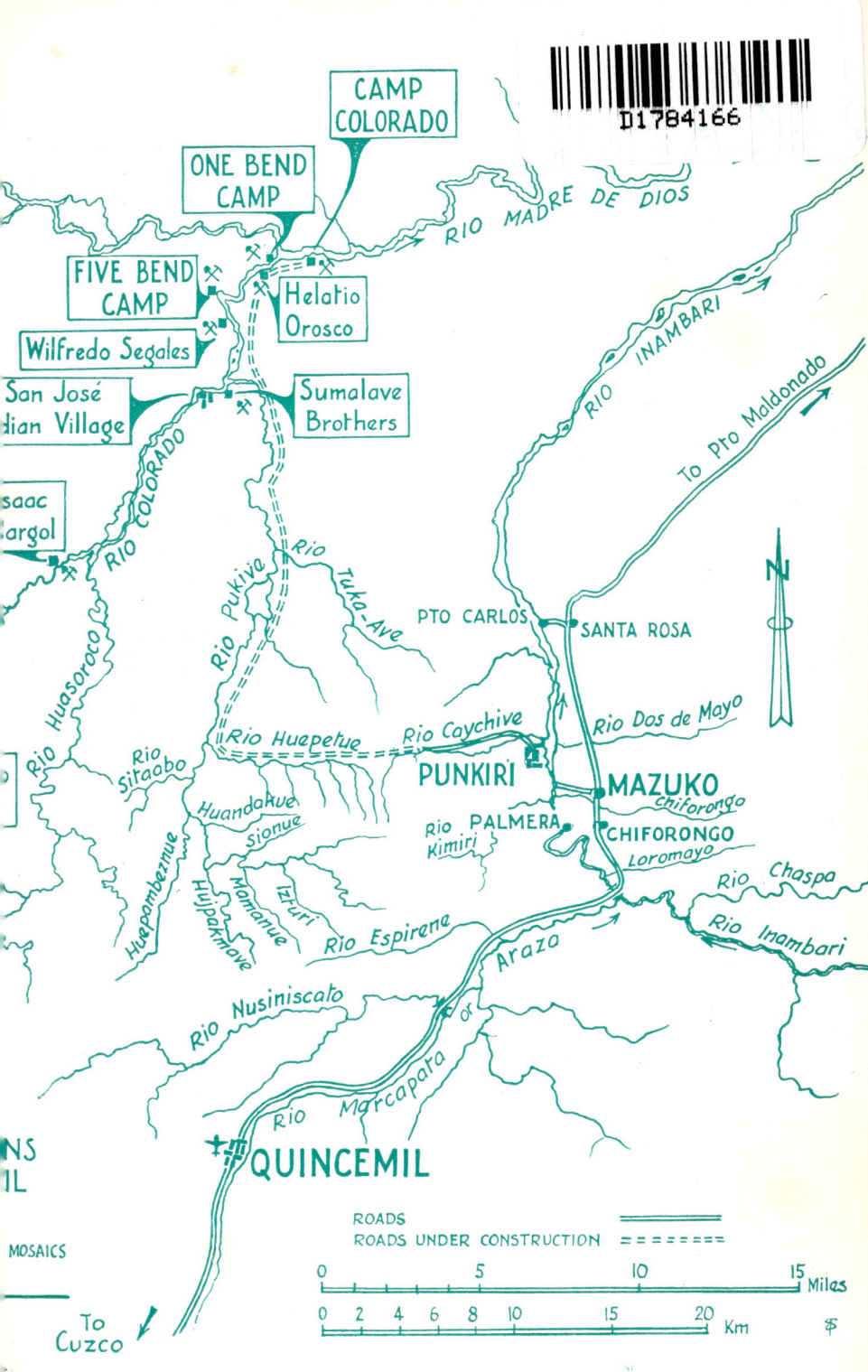

The Gold in the River

ANNA ASHESHOV

The Gold in the River

A Journey in the Jungles of Peru

HODDER AND STOUGHTON
LONDON SYDNEY AUCKLAND TORONTO

To my mother

Acknowledgments

THE DAY THE *Daily Mail* said yes to backing our journey was a red letter day. I had to keep bothering Brian Freemantle, the Foreign Editor, with small queries—but important to us—with which he dealt quickly and cheerfully which helped us get off with a minimum amount of fuss.

The *Peruvian Times*, who know more about jungle travel than almost anyone else, put us in contact with all the organisations in Peru from whom we needed advice and help. They spent a great deal of time making arrangements for us, allowed us to use their facilities, and told us the best approach to any queries we had and generally processed us through Lima in what must be a record of one week.

Banco Minero del Peru put all their facilities at our disposal— transport, channels of communication (radio and postal), they supplied us with maps, machinery, and transport into the heart of the jungle.

Every member of my family helped me to get things together and gave useful suggestions—joining in with a lot of dogwork which builds the solid foundations of a venture: being family they are more apt to be taken for granted but they knew how much I depended on their encouragement.

I had a great deal of help from friends who gave their time, patience and interest during the production of this book. I'd especially like to thank my brother, Peter Frost, and Elsie Herron.

I've often wondered why firms should actually be expected to supply their products free, to a journey of this kind. If they knew of the amount of pleasure travellers gain, first of all from receiving these magnificent presents and then from putting them to the use for which they were designed, though in extreme conditions as we had—maybe they would be encouraged to continue their generosity: it is highly appreciated.

Contents

Illustrations

Illustrations by Katie Bühler and Anna Asheshov

Glossary

Acemilas—pack animals
Ají—hot Peruvian pepper
Cañabrava—wild bamboo
Centavo—cent
Chacra—small farm, plot or smallholding
Chancho—pig
Charapa—turtle
Charpas—nuggets
Choclo—corn-on-the-cob
Chonta—hardest jungle wood
Fariña—dried *yuca*
Huevos de charapa—turtle eggs
Jergón—fer-de-lance
Latigo—whip
Lagarto—alligator
Mñaana—tomorrow
Mantona—snake (boa family)
Makisapa—type of howler monkey
Motorista—boat driver
Papaya—paw paw
Paujíl—wild turkey
Peon—worker
Pueblo—town
Quebrada—creek, stream, canyon
Quepiris—human porters
Rayas—stingrays
Rio—river
Ronsoco—largest living rodent known also as *Capybara*
Sachavaca—tapir
Sabalo—type of big river fish
Serranos—men from the Sierra
Shushupe—bushmaster snake
Tigre—jaguar
Tigrillo—ocelot
Tortilla—omelette
Uncucha—root vegetable
Venado—deer
Yuca—staple root crop used as we use potato

There's gold, and it's haunting and haunting;
   It's luring me on as of old;
Yet it isn't the gold that I'm wanting
   So much as just finding the gold.
It's the great, big, broad land 'way up yonder,
   It's the forests where silence has lease;
It's the beauty that thrills me with wonder,
   It's the stillness that fills me with peace.

from *The Spell of the Yukon*

by Robert Service

CHAPTER ONE

Preparations in London and Lima

"Veins of Solid Gold, many metres thick"

I FIRST HEARD ABOUT the Gold Rush when I was in Lima visiting my brother who runs the *Peruvian Times* group of magazines there. He was talking to a couple of his people in the office about who should go down to look into it and how and when.

"How about me?" I said, which he treated as a not very amusing interruption to the serious business of the day and, as far as I remember, changed the subject.

In the evening I asked him more about the gold. He told me that the main area seemed to be down in the Madre de Dios jungles, near the Bolivian border in the remote south-eastern regions of the country, beyond the Andes. From these areas the Incas and their predecessors had amassed vast quantities of gold which the Spaniards had promptly shipped off home when they conquered Peru four hundred years ago. They had also forced the Indians to work the mines and rivers. Many of these were still known and some were still worked sporadically; but many had been lost or their whereabouts were obscure, old maps marked with crosses giving rise to intense speculation from time to time.

Only a couple of roads, both of them of poor, cart-track standard, went to the area, which was sparsely settled and still

contained a lot of Indians, not all of them friendly. Only two
years previously — in 1970 — the *Peruvian Times*'s Chief Staff
Correspondent, Robert Nichols, had been stoned to death by
Machiguenga Indians in the upper Madre de Dios while looking
for a lost city. I had known Bob Nichols but had not realised
that the area I was planning to visit was so near to where he'd
been killed.

Now that the price of gold was shooting up there were reports
that hundreds of people were going over the mountains and
down into the jungle rivers to find their fortunes. I am not a
journalist, I'm a fashion consultant, but I am keen on travelling
and I thought — well, why shouldn't I go and find my fortune?
I had heard of plenty of worse ideas. Two years earlier I had gone
down the Amazon from top to bottom with John Ridgway and a
couple of others and, allowing for the occasional difficulties and
annoyances, had found that the jungle was a pleasant place to be,
not at all as gruesome as many people suppose.

That evening I told Nick that I was going to try and organise
something. He was in two minds about it — I think secretly he
wanted to go and find his own fortune panning gold instead of
sitting in Lima — but he agreed eventually that it wasn't such a
terrible idea. It was August, though, and the rains in the upper
Amazon start in October, lasting till March or April, so there was
no question of going that year. At the same time this gave me an
opportunity to organise the venture properly.

I started thumbing through anything at all that I could find
about the area. There was very little, and in fact until recently
most people had avoided that region pretty successfully. Among
other things I read that the Rio Colorado, where, though I
didn't know it then, I was to do my gold-hunting, had a special
reputation for dangerous Indians, which had continued well into
the 1950s. Then I came across a few lines written in a dusty
volume of the *Peruvian Times* in 1941:

It was well known that between the Alto Madre de Dios and the Rio Inambari are the large rivers Colorado, Chilive and Blanco. Strange rumours have circulated about these rivers. Tales of fabulous wealth in gold and the incredible cruelty and savagery of the Mashcos. It was reported that the Mashcos are cannibals. Another rumour referred to ancient mines close to the headwaters of the Colorado, where veins of solid gold were said to exist, many metres thick.

Just on the strength of that report, a stockbroker friend later told me, I could easily have floated a company on the Stock Market. Maybe it had sounded particularly good after a few gin and tonics one evening.

By the time I'd put the old volumes back on their shelves I had already decided on one thing: I wanted to find gold. Indeed, I was going to find gold. Back in London I decided on another thing. I didn't want to go alone—I'm not that sort of person—and I started to do some serious thinking. Many enterprises of this kind come unstuck because their members don't get on together and end up by hating each other's guts: I had taken the precaution of reading *The Treasure of the Sierra Madre*. I decided the reason for such failure is often that expeditions are made up of people brought together only by circumstances or for their abilities, not taking into account their characters. They have often hardly even met one another, let alone known each other well, before they set off to spend a long period living in each other's pockets.

My idea was to make up a small group of people I really knew well, and who knew me and my failings, too. Too bad about specialists! We would be a small party of friends. Instead of being miserable while making our fortunes, we would have a good time and still make our fortunes. However, later I was continually to wish that I was an expert in something or other. Almost every day brought forth yet another topic that I longed to know much more about. Metallurgy, botany, zoology, geology, and even astronomy were all to crop up continuously.

The line-up was eventually myself, Katie Bühler and Oliver Hart. Katie is twenty-six, tiny with long, flowing blonde hair. She is a former member of the Swiss ski team and now works as a still photographer and film camera operator. On first appearances, it is not obvious that Katie is very independent since, for a start, she looks like a sixteen-year-old; but she had arrived in England with little money and less English, after having spent a year in Norway. She had one contact address however, and quite soon picked up enough English to build herself a wide circle of friends.

She is meticulous to a degree, but after many gibes (though she probably missed the envy we had for her tidy upbringing) she learned to copy our less rigid life-style. But even today she still presses her blue jeans for a nice firm crease and in the jungle could be seen folding them carefully and putting them under her sleeping bag after she had washed them so that they would be crisp and neat. Katie's mother lives in Wengen and so Katie has skied for most of her life. Even with her slight figure she was well known in the Swiss team for her dare-devil attitude to downhill racing and in fact only a badly broken leg prevented her from following a racing career that could have ended up with many titles and trophies.

Then there was Oliver, twenty-seven, six feet two inches tall, with a blond beard. He's an accountant, though lately he had been spending more time with yachts in different corners of the world than with auditor's books. His father was a colonel and had served in Malaya, Egypt and Switzerland, so even though Oliver's schooling was in Britain he did get around. Oliver has the kind of easy-going but energetic character that makes him fit in well with any situation. He is calm, communicative and amusing, but quiet, and he too is an avid skier and a qualified ski instructor.

I didn't sit down with Katie and Oliver and say, "We've got to solve problems and arguments before we go," but we talked about my idea of going with friends rather than technical wizards, enough for us all to realise we wanted to go on a journey to hunt for gold, and above all to enjoy ourselves and still be friends at the end of it all.

I was to be boss of the party. What that really meant was that in the months ahead, when we came to an unpleasant decision that nobody could make up their minds about, there'd be a chorus from Katie and Olly of, "*You* decide, Anna. You're the boss."

Like Katie and Oliver I ski a lot, and I was in the British Olympic ski team at one time. Now I spend some of my time as skiwear fashion consultant for Pindisports in London.

While Katie did her final film exams and Oliver completed a job he was doing, I set about getting financial backing and supplies for our journey, and amassing equipment.

I was, as I've mentioned, not a complete stranger to jungle conditions. I'd learned enough to get by with while going down the Amazon but I hadn't been to the Madre de Dios nor had I attempted to set up anything more ambitious than a tent, and though I enjoy cooking, I couldn't light much more than a cigarette—certainly not one of those rather frightening pressure stoves, or a camp fire in the jungle. This venture was to be very different in character from the Amazon journey and therefore my tales to prospective benefactors could only touch on the possible trials, tribulations and experiences ahead of us—gold, deep jungle, Indians, river and rapids, snakes, spiders and piranhas; being out of contact for months, and so on.

This is where British Business comes in handy. After weeks of letter writing, telephoning and visiting, we had our basic supplies —food, insect repellent, waterproof watches, camping equipment, medicines, face creams for Katie and me. Most important, we had financial backing from a newspaper. Dermot Purgavie,

New York editor of the *Daily Mail*, told me to contact Brian Freemantle, the Foreign Editor of the *Daily Mail* in London. We had talked to Brian, who soon afterwards contacted us again saying yes, they would back us as much as possible. Like all newspapermen, he wanted immediate forecasts of what was going to happen, like how much gold did we hope to come home with? Twenty thousand? Forty? How could we guarantee finding any? Could we guarantee dangerous Indians, piranhas, snakes and so on?

But now we could get there and back.

Katie and I had one entertaining and very wet Saturday buying equipment. First we set off for the Canoe Centre in Twickenham to get life-jackets. I once spent several hours on the wreckage of a balsa raft snagged on a rock among the rapids of the Rio Tambo, in another part of the Peruvian jungle. These jungle rivers can be merciless, and I didn't care to repeat that experience without a life-jacket.

After deliberating over assorted colours (the green was nice, but it wouldn't show up against the jungle), we settled on vivid orange and blue, and headed straight for Pindisports in Holborn. Oliver was still working and had decided to shop separately.

We spent two hours mooning around the store making lists, getting advice, discussing the pros and cons of a vast assortment of sleeping bags, until we came out laden with anoraks, compasses, camping pots and pans that fitted beautifully inside each other, torches, mugs, sleeping bags, knives, forks and spoons, and a lightweight tent.

It was about this stage, after weeks of arrangements and months of hope, that we really started to get excited and actually believe that the whole thing was on. Were we really going to make it now, all the way to South America? And then into un-known country, part of it not even mapped properly? It seemed that we were.

We went to a lot of trouble to be well organised and this made a considerable difference to our comfort and capability when we finally reached the jungle. Both Oliver and Katie are well-organised people anyway and we eventually managed to be both mobile and fully equipped. We could usually even lay our hands on things when we wanted them. This was largely due to concentrated efforts beforehand both in London and Lima.

The main thing I knew, was to be well protected from water — rain, river and general damp. We had all sorts of waterproof bags and it is impossible to have too many of these, the stronger the better. Good food was another thing I concentrated on.

The "dangers", I knew so little about that I did little about preparing beforehand against them. For us, this turned out to be the right approach. In the jungle the best way to keep out of danger is to avoid it, i.e. *look* before you put your hand into a dark bag (spiders) and kick a log before picking it up for firewood (snakes).

About snakebites, for instance — this turned out to be such a controversial subject that no two "experts" I spoke to gave me the same advice. So, I contacted a friend, Sean McDermott. I got to know Mac on the Ridgway source-of-the-Amazon expedition, when he was the entomologist on the team and was also in charge of our medical supplies and treatment. Now he was working at the Liverpool School of Hygiene and Tropical Medicine, so I tracked him down to ask for his advice on snakebites.

He sent me a leaflet which an expert,* Sir Alistair Reid, had produced that contained some sound and unpanicky advice.

... reassurance is the most important as the danger of snakebite is greatly exaggerated ... aspirin or alcohol in moderation are helpful. The site of the bite should be wiped and covered with a dry dressing. A firm but not tight ligature should be applied just above the bite ... A tourniquet occluding the arterial circulation is *not* advised. The victim should then be

* H. Alistair Reid. *Tropical Doctor* (1972), 2, 159.

sent to hospital. If the snake has been killed it should also be taken to hospital; but generally it should be left alone, since attempts to find or kill it may result in further bites. Generally speaking antivenom has no part in the first-aid treatment of snakebite.

So I decided not to buy serum.

I contacted anyone else I knew who had visited this kind of terrain and probed them for helpful hints.

Tony and Marion Morrison, who have done a lot of writing about Peru, and had made wildlife and other films there for the B.B.C. came to see us and gave us much good advice. They told us how best to take care of our cameras, and Tony whetted our appetites by recounting his own buried treasure and gold stories and finished up by telling us:

"I know where there's a motherlode of gold, but I'm not going to tell you lot where it is because I'm saving it for myself." This made us even more excited and anxious to mark our own cross on a treasure map. Then from someone else I received some unsolicited Good Advice.

"Hope you've had your appendix out."

My heart sank. Perhaps it did sound sensible, but I simply couldn't make myself walk cold-bloodedly into a hospital and get them to operate. So I didn't. Neither did Katie and Oliver, though I passed on the advice.

We had every anti-disease injection and vaccination we could think of: a renewal of our smallpox vaccination of course, plus yellow fever and cholera, and T.A.B. The after-effects were unpleasant; we all felt pretty awful for ten days. In fact Katie's arm became so inflamed that she had to take medicine as an antidote to the allergy. Even these days these precautions are not a waste of time. Juan Mendoza, head of the Peruvian Mining Bank's Gold Department, was later to tell us of a virulent epidemic of hepatitis, possibly some form of yellow fever, which had broken out in part of the Madre de Dios and recently killed several people.

We were sent a list of dried food supplies. We pored over this for ages—after all, we did want to eat well. Reluctantly we had to cut out instant mashed potatoes. Too heavy for the aeroplane flight. But we ordered onions, peas, carrots, mixed vegetables, minced beef, macaroni with cheese sauce, plus some luxuries which H. J. Heinz sent us, chicken curry and beef risotto, as well as Chocolate, Lemon, and Butterscotch Whip Supreme. All this must have taken up one of our baggage allowances on the flight over. Surely we wouldn't get hungry with all this.

Our valuable insect repellent arrived. Earlier, I had received the disturbing news from Peru that the area we were headed for was reputed to be especially badly mosquito-infested. The parcel that came from Shell seemed likely to fill up at least half a suitcase. There were several tins of spray as well as numerous boxes of stick repellent, and a slow burning coil which when lit gave off a pungent smoke to keep away the insects. Looking at the various packs, it seemed so much. Would we really need all that? We certainly would, as we found out pretty soon after arriving.

We had special waterproof rucksacks made up by Crewsaver in Hampshire that fitted on to the normal lightweight frame. These were made of extra heavy plastic sheeting and all the seams were bonded—no chance of water seeping in through the stitching. Soon my flat was so crammed full of packages that it seemed almost like Christmas.

The next big problem was getting it all to South America. Our particular charter flight company didn't seem to worry too much about overweight, but it only went as far as New York. What then? Cross that bridge when we came to it.

So we felt we were set to go.

Everyone had some advice for us. Pessimistic advice too.

"You may be friends, but when you see gold you might get greedy. You'll have to agree about the gold takings before you set off. You should make a contract." It seemed just something extra to try and fit in with the general rush before leaving. Nevertheless, we did decide to do something about it. We just wanted a simple, short letter to say how we would split up our fortunes when we made them. We should have known better. Our lawyers drew up a frightening five-page document full of whereins and whereases, which we all solemnly signed. I don't think any one of the three of us has actually sat down and read the thing to this day.

Departure day approached. The mountains of clutter had to be fitted into impossibly small bags. We had reckoned on travelling exclusively by river once we got to the jungle. We didn't plan on long hikes, and if these were to come about then we would, we hoped, only take a minimum requirement at the appointed time. So we didn't feel we had to start off on short rations.

We must have packed and repacked a dozen times, trying to fit everything in. Sleeping bags were squashed to the size of handkerchiefs. We stuffed our army boots like Christmas stockings. Dried foodstuffs were sealed in double plastic bags — everything else was put in more plastic bags and taped down — we weren't taking any chances of water getting in.

Katie was in charge of films. These were sealed in plastic bags with sachets of silicagel to absorb any moisture; our method of combating the jungle humidity.

Our clothing ideas had been fairly simple. Loose trousers, long-sleeved shirts, several pairs of thick socks, two pairs of basketball shoes and a pair of army boots, and two sweaters. Luckily we each remembered to take a belt. These were to prove invaluable as we were to lose more and more weight as the days passed. In addition we took T-shirts, handkerchiefs and underwear. So we were blissfully uncluttered with clothes. We certainly weren't to have any "What shall I wear today?" problems.

It seems so easy now to write down "Here we are at departure day". But the days leading up to it were a hectic scramble to get everything ready. I am quite good at delegating jobs, and with Oliver getting all the film bags sealed, my brother-in-law arranging for the silicagel sachets to be sewn, a visiting friend getting rid of the cardboard boxes with the dried foods and re-sealing it all in plastic bags, even remembering to put the cooking instructions inside, we managed quite well and ended up at Gatwick airport only a little bit late for a very good family send-off on our first leg to New York.

Our journey in the United States by road to Miami was brief and left little impression, though straight three-laned highways with advertisement boards strewn all through each State were so abundant it was difficult to concentrate on the job in hand — getting to Miami as soon as possible. Oliver was the main driver, with Katie and I doing shorter shifts in between. For our one overnight stop we found a large "family" room in a motel for only eleven dollars. We thought it very cheap. Katie was very enthusiastic about the huge department stores, and Oliver fell in love with a Georgia girl's soft drawl at a hamburger stand. We went through Florida without any hitches and arrived at Miami with plenty of time to spare.

At the airport my nerves started to rattle. We were carrying more than double the weight allowance, and our hand luggage alone would have sent the scales rocketing to dizzy heights. We had three heavy rucksacks with our personal effects, and two more bags containing food. These all needed to be checked in. Then for our hand luggage, we had two heavy bags each, a shoulder bag and a camera case each. But the thought of having to pay for the overweight was just unbearable. Excess baggage charges are absolutely penal, and we were going to need every spare penny we had to finance the gold hunt when we got to Peru.

We had a quick recce to see what the lie of the land was at the South American airline's counter. My courage failed me for a moment.

"It's too early," I said to Katie and Oliver. "We'll wait till there are more people." Finally we dragged the gear over to the counter, after banishing Oliver — we'd noticed the check-in clerk was a good-looking Latin American. We thought things might work out best if Katie and I handled it.

And so they did. The airline man saw we had too much luggage but he was at a disadvantage. Firstly, he couldn't take his eyes off Katie, and secondly he had to tell us that our flight would be four hours late, which meant we had to hang around until 3 a.m. The luggage went through, and we skipped like schoolgirls back to Oliver. We had made it without spending even a dollar. Now there was only the hand luggage to deal with.

Perhaps we should have felt guilty at this flagrant attack on the airline regulations. But I've travelled many times without using up my baggage allowance, so this balanced the account a bit. They never give us a reduction for travelling without bags — I reasoned — so why should they charge us more for overweight this time?

We were in a gay, excited mood, and small things like a four-hour delay didn't daunt us in the least. After all, we were as good as in South America and mañana was another day.

When we did land in Lima we were welcomed by my brother, and a whole row of nieces and a nephew, waving madly over and through the railings. But, we were dismayed to discover, no baggage. Serve us right, one could say. Still, it was very worrying and we spent quite some time form-filling and trying not to wave our arms around like the other exasperated passengers who'd also lost their luggage. It had all been left behind in Miami, they told us, and would arrive mañana. So we had to accept this and just hope for the best. The delay gave us a day to catch up on some sleep, anyway. What would happen if they had lost our luggage just didn't bear thinking about. We couldn't possibly afford to re-equip ourselves, and in any case most of the supplies simply weren't obtainable in Peru. But next day, to our profound relief, it all turned up safe and sound.

The next week was spent preparing to leave for the jungle. We had talks with Juan Mendoza, head of the gold exploration and development department at the Banco Minero del Peru. By law all gold found in Peru, he told us, has to be sold to the Banco Minero, and it turned out that they were concentrating on developing gold production in the Madre de Dios. When we told him our plans, he was immediately eager to help. The bank had a geologist in the area who would give us advice, and they would even lend us a motorised canoe.

This was amazing — better than we had thought possible. Someone was interested in what we were doing, and instead of running into all kinds of obstacles and legal difficulties we were getting nothing but co-operation. We pored over maps, and listened to Juan Mendoza's stories of how impossible the mosquitoes really were.

"I sprayed one once with a direct hit," he recounted, "and it just shrugged its shoulders and buzzed off." He had a twinkle in his eyes as he talked but we were to find that he wasn't exaggerating much.

During that week we bought more equipment, things that we hadn't been able to bring from London: primuses, buckets, ropes, washing powder, sugar and chocolate. Katie had acquired a craving for chocolate which was to carry right through the whole two and a half months of her stay in the jungle. Who ever heard of trying to keep chocolate hard in a steaming tropical jungle!

We had been advised by Banco Minero that we should have some helpers for the first part of our venture and that we should take along an awning for them to sleep under. So we had this made. The Bank also wanted us to make a film for them so that they would then have a record of some of the more remote sections of their territory.

The week flashed by. How any group manages to arrive in a foreign country and set off to unknown parts without having a family to lean on I don't know. Certainly it would have taken at least a month to do all we did in that week if we hadn't been able to pester my brother and his newspaper office colleagues. And Consuelo, my sister-in-law, remained as calm as ever, even after seven days of having mounds of gear strewn all over the house, and three extra people to feed and accommodate. Oliver turned out to be an ace story-reader and paid his way by having all five children in bed on time every night while "Olleee" read them to sleep with Winnie the Pooh.

Then the time came for us to leave. We were anxious, in a way, to get off. But deep down crept in that uncertain, hollow feeling —need we leave right away? We could think of plenty of good reasons not to, and a few more days in comfort would be so nice. After all, how long would it be before we had a comfortable bed to sleep in? When would we next get a nice hot bath?

CHAPTER TWO

Over the Andes to Puerto Maldonado

A blanket of heat

WE WERE OFF TO Puerto Maldonado, a town of about five thousand inhabitants, deep in the jungles of the Madre de Dios region of the Upper Amazon; only thirty miles away from the frontier of Bolivia, and ninety miles away from Brazil. The town was the centre of rubber activity in this part of Peru only seventy years ago during the great Amazon rubber boom of the turn of the century. In those days thousands of jungle Indians were enslaved and put to work tapping the wild rubber trees.

From Maldonado we planned to travel upstream about one hundred and eighty miles on the Rio Madre de Dios to the mouth of the Rio Colorado near to where the Banco Minero had a camp. Beyond that we had no firm plans, except to find gold.

We decided to fly to Maldonado. It was, we discovered, possible to get there by road—just. But normally only freight went that way by lorry.

This perilous—even by Andean standards—road ran from the high Andean city of Cuzco, once the capital of the Inca Empire, over the Andes through a pass reaching over ten thousand feet, and then down through more than two hundred miles of jungle. Often it was impassable, and if it rained while you were half-way along it, you could be stuck for a fortnight. We were anxious to get on.

We had by now acquired some considerable additions to our baggage. We had an inflatable tent, all the buckets, food, pots and pans, lamps, as well as some more equipment we had managed to send on ahead to Peru. We had four hundred pounds not counting our hand luggage, which was mostly camera equipment.

I had decided, rather late in the day, that with all these stories of biting insects we might like somewhere reasonably insect-proof where we could sit down to eat or write in the camp we were to set up. So we had a last-minute sewing session making a giant-sized mosquito net. At home, Consuelo made us inner sleeping bags from some of the family's sheets. We took along spare lengths of mosquito netting material which later on proved really useful for covering clean dishes.

One of the problems of having a lot of equipment is not being able to find something in a hurry when you most need it. We decided now to number each bag and to make lists of contents. This was a long and tedious job which luckily my ten-year-old nephew Igor in Lima took on with some enthusiasm.

All this took time and it was an exhausted household that finally turned out its lights only a few hours before it was time to leave for Lima airport.

Peru is a country geographically split up into three definite regions: coastal desert, the Andean mountain range and the jungle. Very often in the jungle regions large cloud systems build up soon after midday and for this reason most of the non-jet flights crossing the Andes into the jungle from the coast are scheduled to take off in the early morning, so they can return to Lima by early afternoon. (The jets being used these days can usually fly over the weather.)

Our flight was scheduled to leave at 6.30 a.m.—the first of many early morning starts. It was very difficult getting up at

4.30 a.m. We had hardly had time to shut our eyes, after finishing the last awkward bits, like deciding where to keep our passports safely, where we should store our money, and writing last-minute letters to our families.

The money was a problem. Down in the jungle, we were told, it is no use handing over large banknotes because no one ever has any change. We must take as much change as we could manage: five soles, ten soles, fifty and a hundred sol notes, but nothing bigger. Five soles is about the equivalent of five new pence, so as we were taking several hundred pounds to be ready for any emergency, and to use to pay our helpers, we had an enormous packet of banknotes. Katie and Oliver split a bundle the size of a cake tin into four smaller packets and distributed them amongst the bags. We didn't want to lose all our money in one go either.

The final load of our equipment would have nearly filled up a truck. We looked as though we were moving house. My brother said we obviously had plans to stay away a couple of years. Somehow the whole lot, including us, Nick and four of his children, fitted into the car with the boot and roof rack over-flowing. It was still dark and a bit drizzly as the over-laden white Chevrolet lumbered its way slowly across Lima to the airport. Even at that hour there were hundreds of people milling around at the "local flights" end of the airport building. Already late, we barged through the crowd with our cargo.

The girl at the Faucett airline desk didn't bat an eyelid. She calmly conferred with her supervisor and they arranged for part of the load to go some of the way by another flight. It would meet up with us at our second stop—Cuzco. Our flight, they said, was carrying a full load of petrol to the jungle, and they couldn't accept much luggage.

There was a clamour of eleventh-hour urgent requests: "Bring me back a parrot and a monkey!" and "I want some bows and arrows from the Indians!" Then we left the family group waving madly—meeting people at the airport and saying goodbye is still an important source of entertainment in Lima—at the same spot where we had first seen them only a week ago.

We boarded the plane, an old four-engined, propeller-driven DC-6, and settled down. Then we were off on a three-stop flight over the Andes and into the jungle. We were leaving the grey, drizzly winter of Lima to pass through the dry sunny season in the high Andes, then descend to the jungle lowlands, to what they call summer down there, which is really just the drier season. All this in one day. It was the last day of June.

The flight was the best and most spectacular I have ever taken. As soon as we had passed through the winter cloud which sits over Lima, we had a view of the Andes faintly outlined as shadows above the sea of cloud. As we rose, heading east towards the high sierra—the high peaks, 16,000 or more feet above sea level—it seemed as though we could put our hand out to touch the rocky crags and undulating hills. The air became so clear that we could count the sheep grazing on the autumn-brown and yellow coloured slopes. We could see grey stone walls, and tiny adobe dwellings. Our altitude was 18,000 feet—just 2,000 feet above the sierra. It seemed much nearer.

The aeroplane started losing height, and we circled over the town of Ayacucho—an old Spanish colonial town of, I'm told, rare charm and distinction, the centre for a huge area of the south-central Peruvian Andes—before coming in to land. It was a really bumpy touchdown. We only realised why after looking behind as the plane swung around to taxi the few yards to the airport building—a small hut with a huge, old-fashioned fire engine standing impressively beside it. The strip was made of dry, packed earth. As the plane turned, a small group of waiting passengers-to-be disappeared in a thick cloud of choking red dust.

Everyone wandered around informally while a petrol bowser appeared from nowhere to top us up. Then we took off again, circling around between the mountains to gain height and then turning south-east to follow a deep valley with a chain of Andean

(a) *Anna, Oliver and Katie* (b) *Alfonso* (c) *Horacio* (d) *Villanueva*
(e) *Victor with baby tortoise*

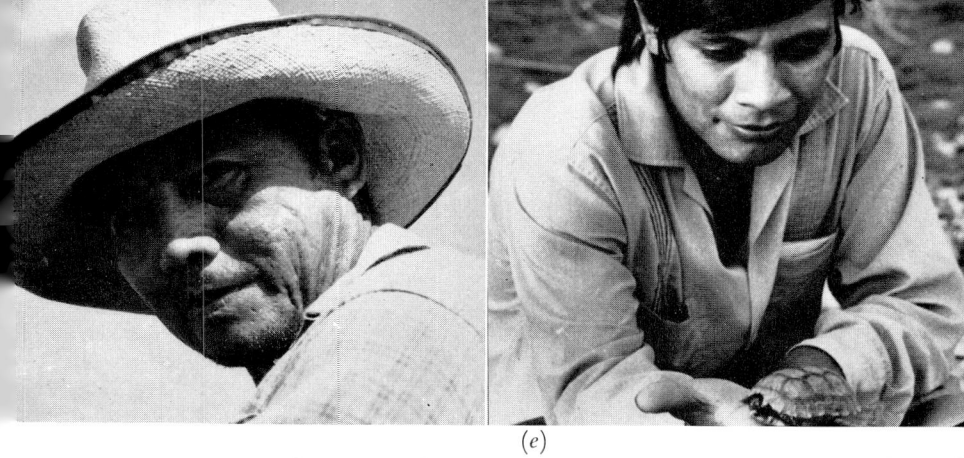

(a)

(c)

(e)

peaks to our left, snow-covered and very beautiful. Two breath-taking mountains seemed to swamp all the rest and the pilot took us right between them. On one side of the plane was Pumasillo—the Puma's Claw—and on the other was Salcantay, one of the highest peaks in Peru at about 22,000 feet. We were flying below its summit—we were able to confirm that from the instrument panel, because the pilot invited us to join him in the cockpit. The exotic panorama of colours ranged from the greys and rusty reds of the barren lower peaks, rich greens and yellows in the cultivated valley below us, and the brilliant blue-white of the snowcaps to the north.

Soon we landed at Cuzco, 11,400 feet above sea level, after what seemed to me to be a kind of bombing run-in up a long and not very wide valley. You could look up and see llamas and alpacas looking disdainfully down at you. Nowadays Cuzco is the main draw for tourists coming to Peru, so the place is becoming quite sophisticated. Small jets land at the concrete airfield, and the airport building looks like a scaled-down version of an inter-national one. We had a nice surprise waiting for us—there was our baggage, piled high on the tarmac; not just a couple of pieces, but the whole lot. It had all come by jet. We were all relieved to see it.

While the aircraft was being refuelled the few passengers remaining with our flight stood around taking in the beautiful sunshine. We chatted with the crew, a friendly bunch, before taking off again, and once more we took advantage of their offer to go into the cockpit after we had passed the first bumpy mountainous section. Before long the peaks and hills began to plunge abruptly into deep valleys and gorges before flattening out into green, green jungle. This was our first view of it, and it was enormous. There was nothing else as far as the eye could see, except this huge expanse of green. Then we passed over several rivers with tributaries stretching away like gnarled old tree branches. The rivers meandered, breaking up the jungle greens

Oliver and Katie—One Bend Camp lunch
Marino finding turtle-eggs

3

with muddy red and grey waters and sometimes golden beaches. We would probably end up searching for gold in one of those areas, we thought.

It was too high to see any signs of life. Perhaps we had hoped to see smoke spiralling up from a hut, or a canoe going downstream, but there was nothing. We could see plenty of ox-bow lakes and meanders, so the river couldn't be too fast-flowing we thought, though this turned out not to be very exact. Our mood was somewhat different from the morning's euphoria over the beautiful mountains and snowy peaks. The jungle looked forbidding.

As we came out on to the plane steps at Puerto Maldonado the blanket of heat hit us. It was like going into a sauna bath, but here you couldn't walk out of it again when you'd had enough. There were small groups of people standing behind the wire fencing at the edge of the airport — dark-skinned pretty girls wearing light, sleeveless cotton dresses, with sandals on their feet, and men dressed in short-sleeved, loose-fitting shirts and baggy light-weight trousers. And here we were, dressed for the jungle — or so we thought — covered from ankle to wrist. There wasn't a mosquito in sight.

Heavily laden and overdressed, we staggered to the shade of the tin-roofed airport building, carrying all our bits and pieces of hand luggage. The airline staff scrutinised the numbers on the tickets of our registered baggage much more strictly than either Heathrow or Kennedy would ever do. And then we were through.

Our movements had slowed down considerably in the heat and we now knew why we most needed our handkerchiefs — to wipe our dripping brows, though within a few minutes we had to start using the backs of our sleeves as well. Standing by the entrance to the airport building was a yellow lorry with Banco Minero del Peru written on the door. This lorry had come to meet us and also to take along some fellow-passengers who turned out to be not gold, but oil prospectors, from Phillips Petroleum. The lorry took us all through the bumpy, dusty streets of Puerto Maldonado, streets lined with thatched wooden huts, and barefoot children waving enthusiastically to us as we roared by. There were some dark little shops too, a market, a couple of cafés, all attractively wooden and with old rusty "Coca-Cola" signs nailed haphazardly above the doors. The lorry drew up on the outskirts at the far side of town in front of a large wooden building raised on stilts, with the window openings covered with wire mesh – "to keep the insects in" as Oliver later remarked, not altogether fairly. This was a government-run hotel where we were to board until we set off upstream.

Down the steps came a smiling young man. With outstretched hand, he introduced himself to us as Benigo Fuentes, though later on we discovered everyone called him Hugo. He was going to be looking after us and seeing that we had all we needed. He was in charge of the Banco Minero's work in this region, and was therefore one of the most important people in town. He had anticipated our wish to get going as soon as possible, and had already contracted six workers for us and sent them on up-river in a motor canoe. We ourselves would be going upstream in a faster boat, and we'd probably overtake them on the way. We were to try our luck working at the mouth of the Rio Colorado, or Red River, a tributary of the much larger Rio Madre de Dios.

Madre de Dios was known by the Indians as Amaru Mayu, River of Lizards, and also as Manutata, Father of Rivers, according to Victor Oppenheim, in his *Exploration East of the High Andes*. And, according to Garcilaso de la Vega, in 1450 the Inca Yupanqui sent an army of 10,000 soldiers into the region. He says 9,000 of the soldiers were killed by the savages, diseases, snakes and other hazards. We ourselves weren't expecting such drastic events.

The heat was appalling and we were completely exhausted after the journey. So they showed us to our rooms and we just collapsed unconscious for a few hours, through the hottest part of the afternoon. There wasn't even a breeze to filter through the mosquito wire on the window frame.

Just before dusk fell we raised ourselves sluggishly, one by one, and freshened up with cool showers. Piped hot water doesn't exist in Maldonado—not that it would be needed—and the town's electricity plant only switches on for five hours in the evenings. We set out to take a stroll round town.

Puerto Maldonado doesn't have many cars. Instead there are hundreds of mopeds. They race around the streets, weaving around the potholes and leaning rakishly into the corners. Often you see young girls speeding along with a couple of even younger children—perhaps their own—clinging gaily on behind.

The other vehicles you see, mainly, are lorries—ones that have made it all the way from Cuzco and even Lima, along hundreds of miles of bad road over the Andes and through the jungle. The journey can take anything from five days to six weeks, the drivers told us. It all depends on rain, landslides, floods, and if you suddenly need a spare part there is unlikely to be an AA box round the nearest palm tree.

We visited the local Banco Minero office which was at that time a tiny clapboard building sandwiched between a café and a grocery shop. A hand-written notice advised us that the price of gold paid is 120 soles (about £1·20 at that time) per gram and it was dated 1st June 1973. There was a set of scales in a glass case on the counter and when we wanted to see GOLD for the first time we were shown the day's takings in a jam-jar with a rusty lid. It felt very heavy but it didn't look much.

Hugo Fuentes smiled at our disappointment.

"Don't forget you will be looking for a precious metal," he reminded us. "You mustn't expect to come back with bucketfuls." Frankly, that's just what we did expect, or at least hoped for.

In a pile on the floor were wooden coolie-hat type, wide, conical bowls about twenty-two inches across.

"Those are the pans you work with. They're made locally from cedar," he told us.

We explained about our four large packets of money. Should we take them with us? He grinned at us.

"You won't have much use for money up there. Take about £20 – S/.2,000 – and leave the rest here. If you need anything we'll buy it for you and send it upstream on one of our cargo canoes." When everyone in the Bank saw our bundles of money there were big smiles all around. It appeared the whole town had been getting very low in small change and though we couldn't finance the whole district it would do for a start. Katie and Oliver had packeted them beautifully for river travel. It took some time to open the plastic bundles. Even if they'd sunk they wouldn't have got wet.

We all went off for a drink, and were joined by another engineer, Victor Vargas. All sorts of unlikely people carry the title "Ingeniero" – engineer – in Peru. Sr. Vargas, for instance was a geologist. He was doing geological research work for the bank in an area not too far from where we were heading ourselves and would look us up from time to time, he told us, to see that we were all right.

We had not expected to find so many shops in Puerto Maldonado, but after looking around we realised we had done the right thing by bringing our own dried foods. The selection of tinned foods ran the whole gamut between tuna fish and sardines, and although we hadn't set off for a gourmet holiday we did want to eat well. When we asked about mosquitoes people told us there were hardly any in Maldonado, which was why everyone could wear short sleeves but we were right to have long sleeves for the Rio Colorado.

"Anna, do you think we need to wear our life-jackets for the river trip tomorrow?" Oliver asked me just before we all turned in.

"Oh, we'll think about it in the morning," I mumbled. I was

too tired to think clearly just then. They were going to call for us at 6.30 a.m. We needed to leave early if we were to make it in one day to Camp Colorado, the new place the Bank was setting up near the mouth of the Rio Colorado.

CHAPTER THREE

180 miles upstream to Camp Colorado

Camping on the river

IN THE MORNING IT was quite clear to me that we should wear our life-jackets — no sense in taking risks. Perhaps I had felt a little shy of being laughed at for overcaution, but when we put them on later they were very much admired, and several people asked if they could buy them.

Our early morning start didn't materialise. It wasn't our fault, or anyone else's. An air force Hercules cargo plane loaded with building materials for the Bank's new offices had arrived and had to be unloaded. As there was only one lorry it had to be used to empty the plane before taking us the ten-minute ride to the Bank's small landing-stage on the Rio Madre de Dios.

This short drive saved us two hours of travelling upstream by river. Puerto Maldonado is built at the confluence of the Madre de Dios and Tambopata rivers. Our hotel, in fact, backed on to the Rio Tambopata. But just upstream of the confluence is a large bottle-shaped bend that goes off many miles to the north before curving around almost back to the town itself. The short, rough road goes through the jungle across the neck of the bottle.

There were three boats at the port. One was a long wooden canoe with a 40 h.p. outboard engine. This was already heavily laden with boxes full of provisions for the Bank's workers upstream. There were a few spaces and in this we stacked every-thing except our camera equipment, a few bare essentials of clothing and food and our sleeping things. The canoe would arrive in a few days, we were told. The other two boats were quite different. They were flat-bottomed aluminium skid boats both with 45 h.p. outboards and we were to load these as evenly as possible so that the two boats would take ourselves and Victor Vargas. There were the two boat drivers, and Oliver and Victor were delegated to fill the petrol tanks from the large drums that were in each boat, so that we wouldn't have to keep stopping to fill up.

We said goodbye to Hugo Fuentes. He said he would be up to visit us some day, and anything we needed just to let him know. How, we weren't sure yet, but it was a kind thought. We were now on the river, leaving behind road, cars, telephones and electricity for quite some time to come.

The journey was hot and tiring, with the jungle sun beating down relentlessly all day long. We were also, we gathered, racing against time because Victor was in a hurry to start work up in the Rio Colorado. So we pushed on, with Victor pointing things out to us every now and then.

Sometimes we would pass wide open beaches—not lovely sandy ones but beaches covered in stones of all sizes.

"That's where you might find some gold," Victor said. We had plenty to learn. It looked barren and uninviting. How could any gold be among those pebbles?

We passed some men working on the beach. Behind them, up on the bank, was a clearing in the jungle. There were a couple of thatched roofs nestled into one corner, and a spiral of smoke circled up, high into the sky.

As our boats raced by it was difficult to see exactly what they were up to, but they were working for gold, Pajarito, the boat driver, or *motorista*, said. We could see somebody shovelling, some water coming from a tube, and a big wooden tray and that was all. Well, we would find out what it was all about soon enough.

All this time both Oliver and Victor in the two boats had been doing a tough and dirty job. About every hour the petrol tanks needed refilling from the enormous drums that sat at the back of each boat. They had a length of rubber tubing and to get the petrol running from the drum to the tank they had to suck, until they got a mouthful of petrol, and then let it run into the tank.

As we'd started so late it seemed like no time at all when we pulled in and Victor announced we were going to camp for the night near a settlement. We climbed up the bank to a group of huts, and set up our inflatable orange tent.

Oliver got the primus going for our evening meal while Katie and I started to sort out the sleeping arrangements. We spread out the sleeping bags on the tent floor, using our opened-out life-jackets as mattresses, but though it was supposed to be a tent for sleeping four it looked cramped just for the three of us.

Until we became more slick in our camping arrangements it was all rather chaotic and finding the insect spray or the salt, or the spoons in the semi-darkness of the tent was a real hassle. Luckily we did have a paraffin lantern which was to prove invaluable. Our dried foods were excellent and just right for this first evening in the wilds.

There were pigs and hens wandering around everywhere, so we weren't keen to leave anything out in the open. Reluctantly we stacked everything into the tent, which barely left us room to turn in our sleep. Victor and the boat drivers had been offered blankets and mosquito nets by people from the huts, so they went off to sleep under one of the thatched roofs.

"Oliver, have you got the spray? There's a mosquito flying right by my nose." Katie was groping in the dark for her torch. Our mosquito netting over the tent opening must have had a hole or a gap in it somewhere. Teething troubles we were beginning to have. Apart from the frequent swipes at the "Zzzzzzz" which kept zipping past our ears, it was unbearably hot. The tent was a heat trap and we obviously had a lot of reorganisation to do for our permanent camp.

None of us slept well, and we were relieved to get going once again soon after day-break.

Later on we stopped once, pulling in at a beach alongside an orange-painted canoe. There were two men who were employees of the Banco Minero, testing the gold-producing potential of various beaches along this stretch of the river. They were delighted to show us how it was done. At the water's edge, one of the men, a tall lanky chap who spoke some English threw a few shovelfuls of the beach into a large wooden pan like the one we'd seen lying on the floor in the offices in Puerto Maldonado, while the other chap, a young engineer, talked earnestly with Victor. We watched as he seemed to swish the water round a few times and somehow made the larger stones and gravel spill out over the edge using a circular movement, and a couple of minutes later our demonstrator stood up cupping some river water in his hand. He gently let this drip through a small heap of about a couple of teaspoonfuls of black sand right in the very centre cone of the pan. As the trickle of water ran over the sand we saw our first gold—as we were ourselves to find it—in its natural state: tiny, tiny flecks of it, not much bigger than the full-stop at the end of this sentence. They glistened with the drops of water running through them. A try-out like that, they explained, in several different areas of the beach, would tell you how many grams of gold per cubic metre you could expect to get from the beach.

We looked blankly at each other. How would we ever manage to tell anything from just those few specks? They, however, seemed to be able to tell just by looking at the sample how much it would produce.

Not long after we passed a canoe, full of men, and stacked high with sacks. A couple of wheelbarrows were balanced precariously on top of everything in the middle. "That's your lot —your men and your equipment," said Victor. "They should pull in to our camp by this evening." We looked curiously at these men who would be working with us over the next weeks, and waved as we passed by.

I'd begun to doze when suddenly:

"We're there," exclaimed Victor, almost out of the blue.

At the river bank stood a brightly painted landing-stage, with a couple of small canoes drawn up alongside and as we pulled in several people came running to greet us.

Camp Colorado was a brand-new settlement of about five buildings which the Banco Minero had started putting up just a few months ago. Maybe one day this clearing in the bush will be marked on the map as a town with a street and a post office. In the future this is to be a main trading station for the higher reaches of the Rio Madre de Dios up to and including the Rio Colorado and its tributaries, all gold-producing areas. They were to set up a shop, a bank office for buying gold, a school for new-comers to the area, to learn about gold prospecting, and housing for engineers and geologists like Victor who needed something more comfortable than a blanket and mosquito net on the ground if they were to live there for many months at a time.

Now there was a scattering of partly finished wooden cabin-like buildings, painted bright blue. Piles of tree trunks lay strewn over the clearing, chickens clucked around tripping everyone over, and there were half a dozen workers hammering, sawing and generally giving a busy atmosphere to the scene.

We had come upstream about a hundred and eighty miles in just over fourteen hours, a journey, though, that would have taken four days in the more usual motor canoe.

Now for the first time it really dawned on us that we were miles out in the middle of nowhere. And soon we would be heading even further up the river, and away from even this small outpost of civilisation.

CHAPTER FOUR

Our first panning

Every speck counts

LATER ON THAT AFTERNOON our six workers arrived with sacks of food for everyone for a month, boxes with tins of tuna fish, milk, and all the equipment we would need for our gold operation.

We were introduced to the stocky, slow-moving man in charge, Señor Villanueva. He had been a prospector for twenty-nine years, and had decided to retire and start up a grocery shop in Puerto Maldonado. Because he was experienced—and because he needed the money—he had been asked to get together a group to come up to the Rio Colorado and help a party of foreigners to get going on gold prospecting. So finally here we all were, eager to get down to work. Victor had told us that although we ourselves would be paying for the men, anything to do with their salaries would be dealt with by the Bank. They'd been afraid that the men would take advantage of our being foreigners —and two of us *señoritas*, at that—and would continually pester me with demands for rises in salary, days off and anything else they could dream up. We were very pleased about this arrangement, as we'd been dreading having to direct six unruly men.

At this meeting we decided against having a communal camp kitchen for the men and ourselves. We didn't want to seem

45

unduly fastidious, but our European stomachs might react violently to bugs that wouldn't even bother those tough men — and they were less concerned than us about the niceties of hygiene.

The Bank put a canoe at our disposal and we set off on a reconnaissance upriver. One of the speedboats went back to Puerto Maldonado and one was kept for Victor's use. We took Villanueva, Pajarito, Victor's boat driver, Victor, and ourselves: a couple of shovels and wooden pans were dumped in the bottom of the canoe. After half an hour we finally turned into a much smaller river, Rio Colorado, 7,000 miles and sixteen days after leaving London. The river could only have been fifty or sixty yards across. Almost immediately there were huge piles of tree trunks, so many that the canoe had to wend its way, first to one side of the stream, in and out, close up to the trunks, and then across to the other side. Often Pajarito would slow the engine right down till we were just nosing gently along. Villanueva stood upright in the bow and with a long pole he measured the depth, showing this to Pajarito, who was sitting in the stern of the thirty-foot canoe, by holding up the end with his hand clutched round the water mark.

Doing this journey time and again we were to find out that there was just one very narrow channel where the boat could pass. If we missed it there were terrible noises as the propeller grated on the river bed, then the shear-pin that acts as a safety device to save the prop from damage would break, and have to be changed. The deeper channel was nearly always at the side of the river nearest to the steepsided banks. The other side would be wide open beaches graduating gently up to small shrubs and bushes before the higher tree line. This, we were told, was the high water mark in winter when the river flooded. Then the beaches would be covered by the rushing tumbling waters coming from the headwaters in the Andes, where fierce seasonal rainstorms were feeding the young streams. As the river meandered all the time it sometimes took up to twenty minutes to get round the widest bends. If we'd walked across the beaches we could have beaten the canoe coming upstream.

Old Villanueva did one or two tryouts on the beach and came back shaking his head.

"I think we can find a better patch," he said. He had a permanently doleful face and I don't ever remember him smiling.

We were to discover that most of these beaches up here did have gold. The point was to find a beach with a good quantity of it. And then, as we were to discover—though not fast enough—it mattered very much what depth you worked at and where. You couldn't pick just anywhere on the beach you fancied.

On we went, further upriver, still passing huge tree trunks; the river level had been going down all week, so they said. This made finding the deepest part more difficult.

We stopped at one or two more beaches and then Villanueva decided that one particular beach would be worth working for a fortnight or so. We were by this time very anxious to have a go at panning ourselves. The first time we'd seen him do it he'd walked to the water's edge and stepped straight in, shoes and all, to a depth of about nine inches.

"Katie, do you really think we have to do that?" I asked anxiously. However, we soon discovered that we were to spend almost all of every day with our feet, socks, shoes, and trouser legs completely soaked, walking in and out of the water most of the day. Indeed, one of our main preoccupations was getting yesterday's footwear dry for the evening so that at least for a couple of hours of the day we could have dry feet.

Oliver was the first to try the panning business. It took him ages. Villanueva would put in a bit of advice now and then:

"Meester. Hold the pan firmly on each side. Use your whole body to make the circular movement. Watch me here, I let the water come in, I swirl it round, it shoots out taking the gravel and sand."

"Why doesn't the gold wash away?"

"The gold is heavy and if you make the proper circular movement it will sink to the bottom in the centre and rest there," he said. Partly he would be explaining away and partly using hand movements to show how to do it. The explanations didn't need translating, it was a matter of copying.

The two of them were squatting in the water swirling away and we could see the huge pile of stones, gravel and sand gradually reduced to a fine black sand in the bottom. All the rest had been thrown off the edges of the pan as the water swirled out. We looked and saw lots of little golden specks mixed with the black sand.

"When we get a production line going with that amount of specks in the tryout, you should get three or four grams per man each day," Villanueva told us.

This, we were to find out, was one of those under-best-conditions-with-best-equipment-with-most-experience, kind of estimates. Six men and the three of us: total, about an ounce a day. A month or two of ounces, and we could fly home first-class and take the winter off. But, as we were soon to find out, that was building golden castles in the air.

This was our first experience of the local brand of rule-of-thumb estimates. It was the same for journeys. We eventually discovered that to ask the question "How long will it take?" was to learn the best time that anyone had ever done under the best possible conditions, for any particular journey. When we didn't arrive even on the day we had expected to we would be somewhat surprised, and it would take us a long time, with many disappointments, before this particular fact of jungle life really sunk in.

However at that point we didn't know, and were eagerly doing some mental arithmetic.

Oliver was just finishing his pan and he had just as many specks as Villanueva. It was Katie's and my turn. Katie had finished almost before I started. She hadn't, though, got the

Washing gravel from the beach

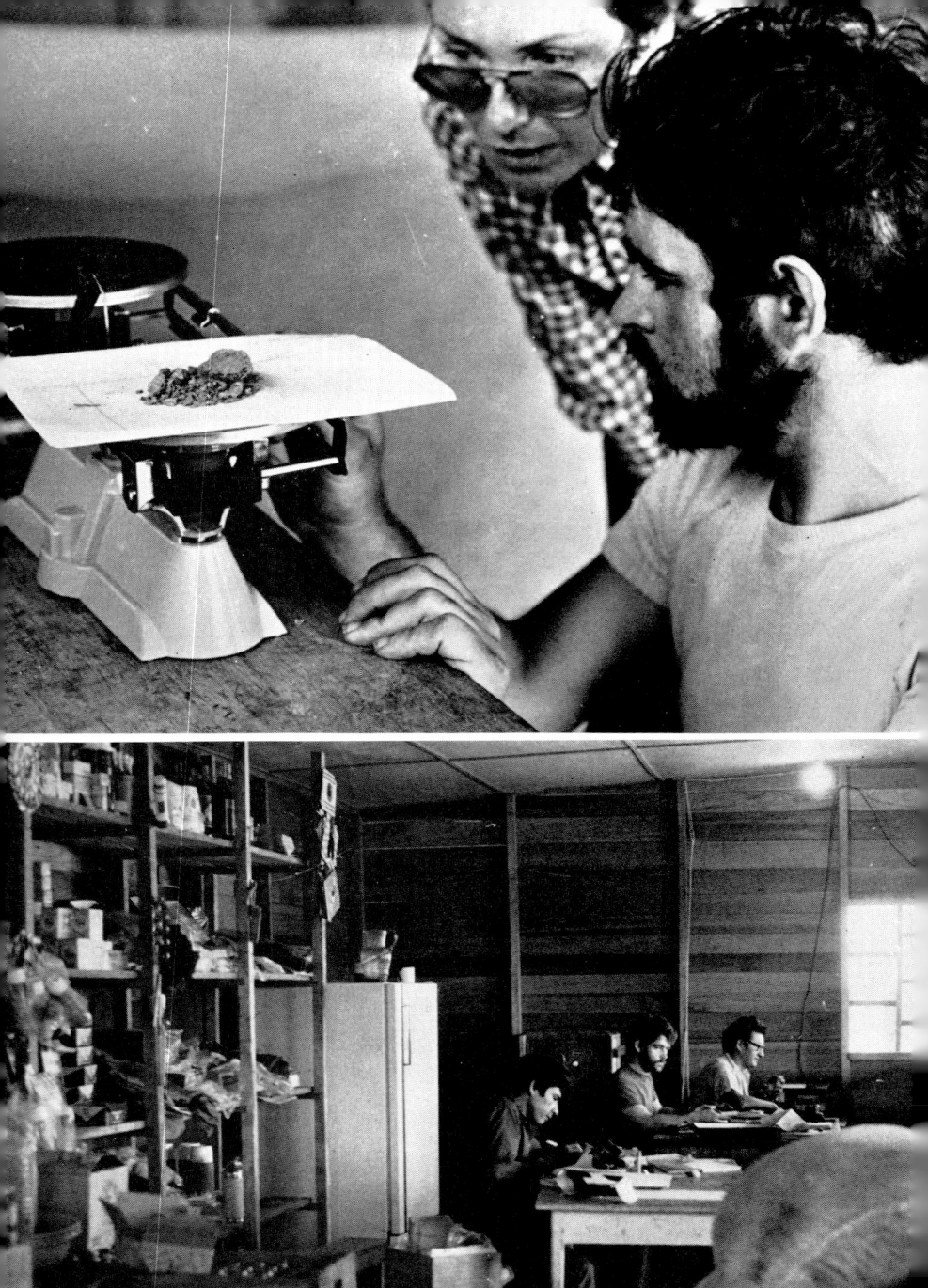

hang of it and in her enthusiasm she'd swooshed everything out of the pan, gold and all.

"*Más tranquílo, señorita*," he told her. More gently—take your time. Many minutes later I'd done my bit, and though it didn't match Oliver's effort there was a bit of stuff in it. We were still at the naïvely enthusiastic stage when every speck counts. We collected the sandy mixture in a plastic bag. Later on we would learn how to extract the gold from it.

The next job was to look for a suitable campsite. We had to be high enough so that if there was a flash flood we wouldn't be washed away. This river had a reputation for turning into a stormy torrent in a matter of just a few hours, though it was almost impossible for us to imagine it that way. Just now it was rather a calm, friendly river.

We all walked around to the back of the beach. Right up at the top where the beach was at its narrowest the ground was higher. Villanueva and Pajarito stamped round it a bit.

"This patch should be all right if we clear the undergrowth away," Villanueva said. "We'll put your section across here, our kitchen here, and alongside will be our tent."

And so it was decided. We would all move up the next day.

Selling our first gold
Camp Colorado office and shop
4

CHAPTER FIVE

Our camp life

Mosquitoes

WE WERE UP WITH the dawn, collecting all our belongings together and seeing that they were well sealed. All this took time, and it was late by local standards—nearly 8.00 a.m.—before we were able to get away.

It was another scorching hot day, and we spent most of it setting up our camp, working with the men. It was a bit of a shambles, since nobody really knew what anyone else was doing. There were no trees this close to the water's edge to use for slinging our hammocks, so the men had to cut nine-foot wooden stakes from the nearby forest and dig holes about four feet deep in which to set them upright. Villanueva, it turned out, had grandiose visions of a palatial shelter for us, and set the stakes side-by-side, nearly six feet apart. We had to use every square inch of plastic sheet that we had to cover this big area, and even then Oliver and myself were too near the edge of the shelter. We didn't like to ask the men to move the stakes after all their efforts of digging them in. We should have been more ruthless: later on when some heavy rainstorms came, both Oliver and I were to get soaked.

Still, for the moment it was pleasant to have plenty of space to move around and the roof was high enough for us not to have to crouch. One of the boys made us a really nice bamboo table, using strips of bark to bind each length of bamboo together. It was such a fine table that we took it upstream with us when we moved to our second camp later on, and we were rather sad when we eventually had to leave it behind.

All round this part of the jungle there's a type of wild bamboo that grows abundantly, called *caña brava*. People use it constantly for putting up temporary huts, to make poles for punting upstream in canoes, and in any structure that needs some kind of support. We had brought our tents not quite knowing whether we would be sleeping in them or in our hammocks. But generally it was far too hot to sleep in a tent, so, except for a covering over our heads, we slept in the open air. The tents, though, were extremely useful and we would have been far less comfortable without them. In the larger tent we kept all our equipment and food. The smaller tent was not used at first, but later on there were, surprisingly, several cold spells, and Oliver would some-times bed down in it.

The men were extremely skilful with their machetes, and watching them construct the frame for our large covering was fascinating. They made two tripods of poles ten feet high, all bound together with strips of bark, and set one up at each end of the shelter. Then they placed a bamboo centre pole about fifteen feet long across the tripods and draped the roofing material across it. We had various different awnings and plastic sheets, so the roofing at the end of the day looked like a patchwork quilt. But it was high enough to give us as much air as we needed, and as long as it didn't rain too hard we were quite comfortable. Our inflatable tent was perfect for changing or washing and for use as a storage cupboard. We had an anti-insect plaque hanging inside which worked wonders, and as the groundsheet was in one piece with the rest of the tent, very few insects could crawl in.

All day everyone worked, levelling out the ground, cutting away the undergrowth. Once Villanueva was attacking a bush vigorously when it literally attacked him. A swarm of wasps shot out of the foliage buzzing furiously and he fled, yelling loudly, thrashing his arms in the air, flicking the insects away from his head and face. Despite this performance he managed to get stung only once, but there were always plenty of wasps around and everyone was to get stung several times during the trip.

When Oliver was digging a trench around our orange tent, he unearthed a bright green and very lively lizard, fatter than the normal ones I had seen before and about nine inches long. When the sun was really hot these lizards would be everywhere. I even reached the stage that I didn't jump nervously when I heard a rustling in the leaves as I walked by a patch of under-growth. It was usually lizards or little birds, though I never stopped long enough to check these noises.

For hours we seemed to have piles of equipment lying around everywhere and it was only towards late afternoon that some semblance of a camp emerged. There was a large shelter on either side of the small clearing, one for ourselves, and one for the men. We'd set up a tent at each end of our shelter, and the men had built a rough kitchen shelter of palm fronds near the river. Finally, there was the light tent of mosquito netting we'd made in Lima before leaving, which we strung up at the back of the clearing, away from the river. There we could relax, talk and eat in the evenings.

To filter our water we had a large canvas bag called a Millbank bag, and Oliver set up a bamboo stake on which to hang it. On the Amazon journey I had been introduced to this type of filter and it seemed quite the best and most portable way to get clean water. The bag is used by the British Army, and I couldn't find any shops that stocked it. After a few enquiries I was promised a couple by the Army for our trip. In return we were going to collect insect specimens for the Royal Army Medical College.

The men's kitchen fire smoked merrily away, and the sounds of wood-chopping, whistling, orders being shouted to different groups, and the thump of spades digging holes for our stakes, all gave an industrious atmosphere to what had been, only twenty-four hours before, a deserted beach.

It was bathing time. We'd done plenty for today, we had a roof over our heads and we had somewhere of our own to sleep. There was a general dash for the river while Katie and I took ourselves off much farther downstream. The beach we chose could hardly be called private, since it was wide open, but as we were to find it best to wear at least one layer of our clothes for our dip as protection against the insects, it didn't really matter. It was just more pleasant to be away on our own.

We set off with our pile of clean clothes, soap, a bucket, foot powder and insect repellent. The river was a bit muddy, fast running and very shallow—only thigh-deep. Katie cheerfully went out near to the middle, splashing around in the cool, refreshing waters.

Had we wanted, we could probably have forded the river. The disadvantage of it being shallow was that the mosquitoes could get whatever part of you that was above water, but there was some consolation in that I didn't expect any big slimy fish to be swimming around. The beach was so stony that I kept my gym shoes and socks on, and dipped under to cover myself in soap at record speed. Swimming in muddy water and thinking I could feel little fish nibbling at my legs did not encourage me to linger. The feeling of dry, clean clothes and dry feet was a luxury we longed for each day, but the effort of getting to that state with both the burning sun and the insects to cope with, was considerable.

We had some air-dried, pre-packed meals which turned out to be a real treat. Tonight, as a celebration for our first evening out in our own camp, we heated up our tinfoil-wrapped Heinz Chicken Curry. It was so delicious we couldn't believe it—way better than our usual fare; good enough, in fact, to serve at any dinner party. Too bad we only had a few packets, so that we had to keep them for special occasions. Until they were finished, though, it wasn't difficult to dream up special occasions.

Everything this first evening was so new and strange that the atmosphere was slightly uneasy. We were finding out about the hardships of camp life, we were sensing each other's doubts, mostly about how we would get on with the men, and we were wondering whether it would really be bearable to sleep out as planned and if perhaps the insects and discomfort would in the end get the better of us.

We were, too, uncertain as to how we should organise the food the men had brought along for all of us to eat. I decided not to do anything, and left the stores in charge of Villanueva — a decision we were to suffer from later on. I also left Villanueva to give orders to the men and see them carried out; I had expected this to be his job. It was only a couple of weeks later that I could look back and see that if I had started ruling the roost firmly from Day One, things would have run more smoothly.

During the weeks to follow, as it gradually dawned on us that the organisation of the men was very weak, I began to put my oar in, and this created bad feeling with Villanueva. Although he was not doing much himself he didn't like to be by-passed either, however tactful and smilingly polite I thought I was being. Today, though, everyone had been working very hard.

After cooking supper among the gnat-clouds outside the three of us escaped into our refuge of mosquito netting. It was terribly hot inside but almost insect-free, so we were able to enjoy a fairly comfortable meal.

"Does anyone want coffee?" Oliver asked. Once we'd climbed inside it was a real effort to get out and start that performance with the primus all over again so we could have a hot drink. There was still a slightly gloomy atmosphere, all three of us wondering what we had let ourselves in for, now that we were actually here; but Oliver's coffee cheered us up a bit.

We chatted a little, munching away at some chocolate that had survived the journey without melting too much, then Katie and Olly decided to go to bed. I stayed to write for a while. There were sounds of teeth cleaning, and shafts of torchlight moved around. Then came a thud.

"What's happened?"

"I didn't make it into my hammock the first time," said Katie, brushing down her jeans and looking up at me sheepishly.

"Look, you take a firm grip on each side and then sit down in the middle and swing your legs in." I already knew about hammocks and I've always liked them. Katie and Oliver weren't so sold on the idea but I'd tried hard to convince them. It was well away from the ground, I'd said: you swing gently in the breeze: the air was fresher higher up off the ground.

I had got the hammocks well in advance of our departure date, and I'd suggested to both Katie and Oliver that they should sleep a night outside in England before we set off. Neither had actually got round to it, which was a pity because it turned out that they both hated hammocks. Katie wasn't really used to sleeping on her back and liked to turn over a lot, which meant she was swinging precariously and nearly looped the loop more than once, waking uncomfortably each time. It was too cold for her back anyway, she said, and emigrated for one night to the small blue tent. Then she was back again the next night. It had been too hot inside.

Oliver, too, had his problems. The hammock wasn't big enough for him and he also likes to turn over, so he made himself a bed of leaves as the men had done for themselves, then he put down a ground sheet, hung his mosquito net over it and slept on the ground.

I slept like an Egyptian mummy, wrapped in my sleeping bag, with my arms fitting tightly inside too. The problem with these hammocks was their narrowness. If an arm accidentally leant against the netting the mosquitoes took full blood-sucking

advantage and we would wake with nasty red patches of concentrated bites. On the really hot nights we just slept in our inner
sleeping bags.

Camp life slowly evolved into a pattern over the following
weeks. Oliver looked after our water supply, cleaning the water
filter and seeing there was plenty of clean water all the time. He
did the repair jobs, like mending the inflatable tent which had
deflated, and helped Katie and me to adjust our hammock ropes
which were often in need of lowering or raising. He usually
lit the primuses, and saw that they were working properly and he
kept the lanterns filled and trimmed.

We had a heap of washing to do every day and there was nearly
always something to mend. Oliver had some favourite jeans which
were more patches than jeans, and these, with the washing
and the work, soon needed patches on the patches. I had been
very proud of my two pairs of trousers, bought cheaply in the
King's Road, only to discover cheapness didn't pay, as both the
zips went early on, and I exhausted our stock of spare buttons
right at the very beginning.

We were woken after our first night by the sound of the men
chopping wood for their kitchen fire. It was still dark. Maybe it
was time for me to get up and start breakfast, but three inches
from my nose my torch beam picked out thousands of mosquitoes stuck to the other side of the netting just waiting to devour
me. I couldn't face it. Their high-pitched whining and the sheer
multitude of them pushing their horrid needle noses through my
netting was too much till it became really light at about 6.00 a.m.
and I just had to get up. Usually mosquitoes clear off when the
sun comes up and it starts to get hot, but in this area, for some
reason, there were great clouds of them around all day even in
the blazing heat of early afternoon.

This was our first morning working, and we were anxious
to be out there on the beach to join in and see what washing for
gold was all about. Around 7.00 a.m. the men all filed off to the

beach, and we followed. The equipment lent to us by the Banco Minero consisted of tubes of plastic, a small petrol-engined water pump, a wooden trough to wash the gravel in, a metal sheet full of big holes to act as a rough sieve, some hessian for laying in the trough to catch the gold, wheelbarrows, wooden planks, buckets and shovels.

Villanueva had selected a patch on the beach where we were to work, on the basis of the tryouts we'd made two days ago. It took about an hour to rig up all the equipment, and when we'd finished we had a kind of Heath Robinson contraption—a long sloping trough with a big box at one end to fill up with sand, gravel stones, and—we hoped—a little gold, dug from the surrounding area of beach; and a flexible pipe to play water over the contents of the box and wash the fine stuff through the holes in the metal tray. There were many teething troubles. The water pump was too powerful and washed all the gold away: the trough was set at too steep an angle, and so on. For most of the morning we stopped and started many times.

We shovelled, washed and cleared away the stones till noon, sweat pouring down everyone's faces. Periodically someone would duck under the water pump for a cooler, and the men had brought a home-made mug to drink from. It was an old con-densed-milk tin with the sharp edges rounded off. The noise of the water pouring from the pipes and the load of gravel being poured into the trough, the spasmodic shovelling away of the wet stones and the chugging of the water pump took on a mono-tonous rhythm. Shovelling away the wet washed stones was my job today. The person who did this was called the *ganchero*.

All the large stones needed clearing, and the stones kept piling up so much that I couldn't stop shovelling and blisters appeared soon after I started.

Lunchbreak came. It was so welcome we downed tools and immediately went for our daily swim, with the hot sun right overhead and beating down relentlessly. Immediately after coming out the air seemed cool — but only for the first instant. In no time we felt as hot and sticky as ever.

Lethargy was a permanent problem in that heat. Will-power dissolved, and you felt like just collapsing in your hammock and staying there. But that didn't help, because it was the kind of heat that didn't make you feel any hotter if you were moving around than if you were flaked out, motionless. And the less you were moving, the worse you were bothered by the insects.

Katie and I were so done in from the morning's work that we decided after lunch to have our siesta in our hammocks. No sooner had we settled ourselves than old Villanueva came over to us.

"Señorita Anna, Marino and I are going to cut through the jungle just near here because from the river we've seen there is a dried-up stream which we want to inspect for gold. We won't be long," he added.

I was already starting to get up before he'd finished talking, and Katie's hammock was swinging as she rose. "We'll come along too."

"But of course. It isn't very far. Would Meester (as they called Oliver) like to come too?" Oliver didn't want to come. He was sunk deeply in *Beneath the Wheel* by Herman Hesse and nothing could drag him from under it.

Marino, the youngest of our workers, aged twenty, was in his bare feet we noticed, but Katie and I busily tied on our army boots and tucked our trousers legs firmly into our heavy socks. We each took our machetes and followed the two straight into the forest. It wasn't thick at all just here and the sun burst through the high treetops leaving pools upon pools of golden light broken by dark patches of undergrowth and dried up leaves and branches. It was much cooler and there were fewer insects: it was good to be going into the jungle itself for the first time. There were many birds singing, and had it not been for Villanueva with his gun slung across his back, and Marino carrying the wooden pan and shovel, we could have been walking in the New Forest in summer. The rocky path was short and we soon arrived at the rocky, dried-up river bed. The brightness of these stones dazzled us and the pebbles themselves were so hot it was difficult for Marino to walk with his bare feet. The men made several tryouts, panning away in the hot sun, and one small section proved to be very good. As we were digging, there was a rustling in the bank and a large black bird with a red head flew clumsily to within ten yards of our little group and settled in a tree. There was a general movement around and as the gun was leaning on an old tree trunk some distance away, I called out stupidly, "Do you want the gun?" However, the bird was even more stupid and didn't move despite my noise.

Indeed, Villanueva was already retrieving the gun and lining it up to get a decent shot at the bird. Marino had taken off at a breathtakingly fast sprint and reappeared only a few minutes later with another gun. But by this time Villanueva had shot and retrieved the bird, which he said was a *paujíl* — a wild turkey. It was big and there were smiles all round. "We will have a good meal this evening. *Muy rico* — delicious," they said.

After this escapade, having missed our siesta, Katie and I opted out of shovelling in the afternoon. We were to do this often, but to ease my own conscience that I should be working, I would find various little jobs to keep me busy. My camera was always a big excuse, the men loved having their photographs taken. Then there was the job of sampling the beach, putting a couple of shovelfuls into a pan and testing to see if the spot where the men were digging was still producing a reasonable amount of gold.

Oliver was far and away the toughest of us three. He was always doing one or other of the gold jobs for the full session, whereas I might opt out before the end, muttering some excuse about putting lunch on. Katie couldn't manage much of the heavy physical work, and she decided she would do the housekeeping and cooking on the days when I was working, so we had the added luxury of coming back from work to find a hot meal waiting.

Twice a day we would peel away the hessian strip from our trough and wash it out thoroughly into a large petrol drum we kept for the purpose. We'd also scrape out the sand from the bottom of the trough into the same drum. Each time we did this there'd be a bucketful or more of mixed yellow and black sand in the drum, but we couldn't see any gold yet. Villanueva said that we'd extract the gold from it after a week's work.

We were already getting bored with our diet, which consisted mainly of dehydrated packet foods plus rice. Soon everyone was craving for a nice crisp apple, or a lovely bowl of lettuce and cucumber with tomatoes. Katie's special longing was for a big plateful of Swiss cheese with some sausage and a huge crust of fresh bread with butter. These longings were hopeless, of course, as none of these was even remotely available. Neither were other city-life staples, like fresh milk, cream, butter, potatoes or any fresh green vegetables.

In this particularly hot period which was happily not to last forever, I could get up little enthusiasm for food, anyway. The only things I could get excited about were cups and cups of tea to quench my continual thirst. The extra London pounds peeled off, which did really please me.

Oliver, though, was hungry the whole time, and his appetite so good that Katie and I often badly misjudged cooking amounts because we just couldn't visualise so much being eaten. The men ate plates piled high enough for three normal servings, and then asked for second helpings.

Our wild turkey was—as Villanueva had predicted—delicious, though a little tough. The men cooked it, and we gave them some of our dried beans and carrots to add a little colour to the stew and a handful of dried onions which gave it an excellent flavour. After supper one of the young boys, Lucho, went fishing with a line and a huge hook, taking the turkey's entrails to use as bait. Later he came back with a fish nearly a yard long — some kind of catfish, with long whiskers and a vicious-looking mouth. This was to be next day's breakfast.

CHAPTER SIX

Operation gold

Deer and a snake

OUR GOLDING OPERATION WAS turning out to be completely different from what we had originally expected. We thought that we ourselves would be squatting down and panning all the time, as in films, but we discovered that you only pan gold when looking for the larger lumps, found higher up near the river headwaters where we were to explore only later on. Meanwhile we were busy learning how to make our present method work most efficiently, since we were very keen on finding enough gold to pay at least for the men and all the food supplies.

The gold we were extracting from this beach came in tiny flakes. There were plenty of them, but then to make up any weight at all there needed to be plenty. Back in Puerto Maldonado our men had built us the conventional type of wooden channel used by everyone in the district. It was built of cedarwood, about nine feet six inches long and three feet wide, with edges five inches high. At the washing end, the edges were built up much higher, forming a ledge running round inside and at the end, to support the sieve. This was a metal tray made from a flattened oil drum, and punched full of slits, which allowed sand and fine gravel

GOLD WASHING ON A JUNGLE RIVER

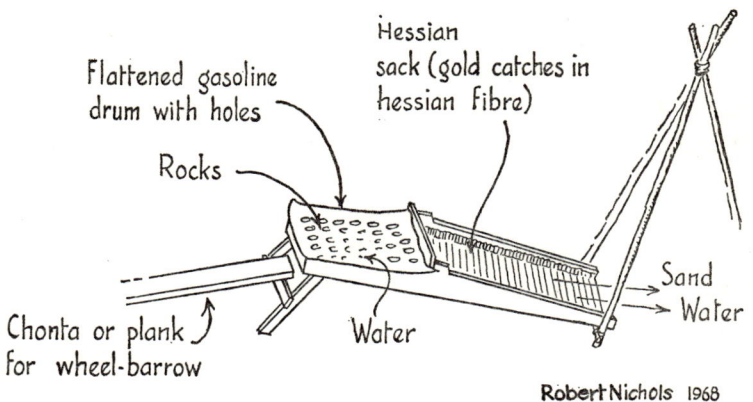

Flattened gasoline drum with holes

Hessian sack (gold catches in hessian fibre)

Rocks

Sand

Water

Chonta or plank
for wheel-barrow

Water

Robert Nichols 1968

through, but caught all the large stones, which were then promptly thrown away after washing them.

Every day we would lay hessian into the bottom of the trough. Four of us would dig the stones and sand from the beach, filling the two wheelbarrows though chucking out the biggest stones. The two strongest men would push the wheelbarrows up the planks and tip the contents into the trough. On one side stood the washer holding the end of the tubing from which the water poured. He played the water on to the stones while the sixth person standing opposite, turned the stones over using two small pieces of wood six inches by four. When clean, the stones were pushed out a few at a time. Meanwhile another job was that of shovelling away the pile of washed stones so that the trough wasn't buried.

All this time gravel, sand, black sand (iron ore) and gold,

filtered through and on to the hessian with the heavier black sand and gold catching in the weave of the hessian and the rest of the material washed away down the trough and on to the beach. This whole system worked, provided the water flow was not too strong and the trough was set on its bamboo tripod stand at the correct angle of fourteen degrees. Twice a day we would rinse our hessian out into the big oil drum and retrieve the remaining sand and gold particles stuck to the trough by carefully pouring a bucket of water down and catching the run-off in a couple of pans.

But that was the fun part, cleaning out the trough and seeing how much we had. Most of the time it was just a hard slog. Where the men had been digging there was a dip in an area of the beach about a foot deep.

Every day we had to shift to a new spot, for the men dug over an area of about twenty square yards per day and it was a waste of effort for them to have to push the wheelbarrows too far. Setting up the whole system took about an hour each morning. The water-pump engine was at the water's edge and plastic tubes, each over three yards in length, joined up to make a pipeline of sometimes as much as ninety yards. As the piping neared the trough it had to be raised gradually to stand above the trough, and this was accomplished with the versatile bamboo tripods. Setting up the plank was often the most difficult task, since if it was too steeply set the men just could not push the wheelbarrow up the last few feet of the plank to tip the stones up and over into the trough. Once or twice there were casualties when the wheelbarrow tipped over on its way up the plank and missed the trough completely, much to everyone's dismay. It had taken a lot of effort, after all, to fill the barrow in the first place.

When we had enough sandy mixture in the big drum, after about a week's work, it was time to clarify it. That, we discovered, was the technical term for getting out the gold.

What we had in the drum was sand, black sand, and gold

particles. The next stage was to pan away the sand, leaving only the black sand and gold. This was a job for the experts – beginners could easily muff it and wash out precious gold – and Villanueva started off working on it with Alfonso, our *motorista*, who was also a very experienced gold prospector. Oliver soon became very good, though, and later joined in this part of the operation. They had to pan away gently, methodically and rhythmically, never getting too violent a movement going, swishing the light sand away over the edges of the pan. It was a more skilled operation altogether than the tryout panning that we'd been doing with the gravel and large stones. The heavy black sand and the gold left behind were then put into a bucket, and when we had more than a bucketful (this was so dense and heavy I couldn't lift it by myself) it was time to extract the gold from the black sand.

For this we needed mercury, and we had with us a small bottle – which seemed to weigh a ton – of this incredibly dense liquid metal. We watched Villanueva and Alfonso carefully. They took a couple of handfuls of the black sand, put it into the pan, added a few drops of mercury and started kneading the mixture in much the same style as one would knead dough. They said the best method was to knead it with the base of the palm. The principle behind this operation is that the mercury amalgamates with the gold (but not the iron ore black sands) – "sticking" to it as it were. This amalgam is much heavier and denser than the black sand and so when it's all swished around in water in the pan, it sinks to the bottom so that the black sand on top can be swished out of the pan, leaving the gold mercury mixture behind.

When they could see, on close inspection, that the shiny specks of gold were now a silvery colour, they were ready for the next stage. Down to the river's edge again, and the men started the panning movement, very very gently shaking the pan from side to side so that the gold would sink to the bottom, and then, still very cautiously, making a wide circular movement, swirling the water slowly around. Gradually the black sand disappeared over the edges of the pan until only a small, dull, silvery mass remained. They put this on to a plate, and started the process all over again with a new lot of black sand till the bucket was empty. This silvery mass was the amalgamated gold and mercury. Now it was time to burn off the mercury and see what we had travelled more than seven thousand miles for—river gold, 96 per cent pure, or 23·5 carats.

Villanueva hunted round for an old fruit-salad tin, washed it out well, took off the lid, and got hold of an old bit of damp rag. Then he got a wood fire going to a really good blaze, choosing particular types of wood that burn with the hottest flame. After putting a piece of paper in the bottom of the tin so the gold would not stick, he tipped in the silvery mass, laid the damp rag over the can and started heating it.

"The rag is to catch the mercury as it vaporises, so that we can use it again," said Villanueva.

We kept our distance. Mercury vapour is a powerful poison. We didn't know how much you needed to inhale before it really got you, but we weren't taking any chances.

After peering inside several times Villanueva finally pronounced that the gold was "cooked". He set the tin down to cool for a while before turning it over on to a plate. There was this lumpy mass of bright yellow material. Gold. It couldn't have been mistaken anywhere for anything else.

So there was our whole process. We had only to continue the hot, strenuous slog to get a pile big enough to make us feel that we hadn't come all this way for nothing.

Back in Lima we had been told by the Banco Minero that they suspected there might be platinum as well as gold in the Madre de Dios area. They asked us not to mention it though, as they were in the early stages of testing samples, and it was far too soon to raise the hopes of the miners. But platinum! It was worth far more than gold! We used to refer to it secretively as

"P" and suffer terrible anxiety twinges thinking we were panning away a fortune by not knowing how to extract it.

"Look at all that P we're losing," I'd whisper to Oliver,
showing him a handful of black sand full of shiny, silvery specks.

Ultimately we heard it was all a false alarm, for there was no
platinum in the Madre de Dios. The shiny dust turned out to be
worthless mica.

We worked hard for many days. Sometimes we would switch
jobs to tax different sets of muscles and make life more interesting. One day I would wash, and another day it would be Katie.
Oliver usually turned over the stones, but sometimes he would do
tryouts—"assays", in mining parlance—on the beach all day,
because we had noticed that the men were inclined to dig haphazardly anywhere and to any depth. It hadn't occurred to us for
the first days how important it was to keep checking both depth
and area. We thought old Villanueva should have been over-
seeing that part of the job.

I asked Villanueva if it wouldn't be a good idea to have a go
up in the little dried-up fork of the river he had discovered the
other day, and work that for a short time. It didn't seem as if our
own production was anything like as prolific as we'd been led to
expect. It was only then that he told us a couple of men were
working up there already so we couldn't go. This was our first
experience of missing out on a good patch by not moving in the
very moment we discovered it. I still wonder which of our chaps
let on that we had found a good spot to that pair of men, who
were to be our neighbours for the rest of our stay at this
camp.

We would go for days without seeing or hearing anyone, and
then two or three canoes would pass up or downstream all in
one day. Normally, nothing would ever slip by unnoticed. It
was a good excuse to stop work for five minutes and discuss who
it was, or might be, that went by, where they might be coming
from, going to, why, and for how long. If the canoe had a motor,

then the discussion started long before they came into sight, because the men knew exactly which kind of motor had which sound, and more than likely they knew who owned that particular motor. We saw an anti-malaria boat from the health authorities go by. This campaign, under way for twenty-five years and more, has been highly successful in Peru and an outbreak is now almost unheard of anywhere in the vast Peruvian jungle — or, for that matter, in the coastal valleys, where the disease used to be endemic. They visit everyone's house or hut and spray with insecticides. Occasionally there were traders going upstream to sell their wares. We heard that these people were real profiteers who sold their goods at exorbitant prices, usually in exchange for gold. Sometimes we'd see woodcutters floating downstream, headed for Puerto Maldonado on their homemade rafts, and every so often Victor Vargas would come to visit us, as his work in the jungle was not so far away.

One day Victor came quite late in the afternoon, saying that he could only stay for a few minutes as he had to get back for the daily radio transmission. The recently installed radio was now working — sometimes.

"I just came up to tell you that Hugo Fuentes sent a message for you last night over the radio. He's sending up a bag and a package that arrived for you from Lima."

"Whoopee! When will it get here?" It was my birthday next day, and I still had a kind of day dream that I would get some birthday post.

"Oh, probably in about six days' time," he said. That dashed my hopes a little, but it had been too much to expect anything, anyway. Katie, too, had been hoping for some letters from her fiancé, and we were all beginning to feel the effects of isolation, in the shape of a craving for news from the outside world.

As Victor waved goodbye in his dash to beat both dusk and the clock he called out:

"See you tomorrow. It's Sunday, so we can all go fishing and explore the jungle a bit." And off he went down to his camp, which now seemed an awful long way away. Our own little encampment had become our whole world. There was nothing to remind us that the Banco Minero's camp was only a few miles away, and any visitor seemed automatically to have come from a great distance.

That evening as Katie and I were in our hammocks the trees were swaying with a breeze that had sprung up. Suddenly there was a curious rustling noise, as if a bird or animal were pushing its way through the undergrowth.

"What's that?" said Katie. She struggled into a sitting position and shone her torch into the darkness. I could hear the tone of alarm in her voice.

"I thought it was the trees rustling in the wind," I said.

"No, no. It was something quite big. What shall we do?"

I climbed out of my hammock and shone the torch around nervously. Maybe I would scare it away. Or would I see a pair of big red eyes staring me in the face? There was an abrupt thrashing in the undergrowth, a few feet away and something went charging off into the forest, butting and crashing through the thickets.

I had to sit down very gently in my hammock and let my pulse rate drop back to normal. Actually, to be rational about it — which wasn't possible at the time — if it *had* been anything dangerous we would almost certainly never have heard it. Scorpions and tarantulas are too small to make a noise, snakes slither soundlessly, and jaguars or pumas are as silent as ghosts. Probably the animal was just a harmless old *ronsoco*, a huge rodent that lives in riverbanks all over the jungle.

Even without real scares, the night sounds of the forest gave plenty of food for the imagination. Animals chattered and howled in the darkness, birds called to one another, the river rushed over the rocks close to the camp, and the rapids further downstream kept up a constant dull roaring. Often part of the river

bank would crash into the fast flowing waters, making a tremendous noise. Then the crickets and cicadas would start up, whirring
and chirping. It was enough to keep any nervous person awake.
Normally, though, we were much too exhausted to worry, and
only really loud noises made us jump. But with the feeling of
security from my hammock I loved these sounds at night and
only wished I was a proficient zoologist so that I could identify
the jungle orchestra.

Even in the jungle it's possible to have a lie-in on Sunday,
though without curtains it seems a little ridiculous to stay in bed
with the sun shining on the camp soon after six-thirty. Today, in
any case, we were going hunting and fishing to get something for
my birthday dinner.

I was longing to catch one of those huge fish, so I asked to
borrow one of the hooks and lines that the men had. I would
never have dreamt of bringing anything like these things from
England. The deep sea type hooks were enormous—at least four
inches long—and you could have fitted an old penny piece into
the curved part. The line was thick nylon cord and this was
wound round a piece of wood. Marino baited my hook for me
with a huge handful of guts from another wild turkey they'd
shot.

"*Señorita*, if you get a bite, wait a little until it *really* bites.
Then pull with all your strength, because if a big fish bites they
are very strong—maybe stronger than you. Be careful."

I set off down the beach as far as I could go. It must have been
nearly a mile, but try as I might, I couldn't find a deep pool. The
sun was doing its level best to shrivel me up again, so after trying
to throw the line far out into the river many times I trudged back,
somewhat discouraged and quite ready to go hunting. As I came
into camp there was not one of the men in sight, and the canoe
had gone.

"Where are they, Katie?"

"I don't know, they just took off without saying anything.

They all went and they've got both the guns with them. They headed upstream."

I was furious, and worried. They hadn't asked my permission. We had been lent the canoe, and so we were responsible for it, and if anything happened to it I just couldn't imagine all the consequences. Anyway, we might have wanted the canoe for ourselves to cross the river to go hunting or find deep water pools for fishing. This was not a very good start to my birthday.

Soon Victor arrived. He was quite as cross as I was that they had gone off, but it was dangerous for him to take his boat further upstream. The propeller went very deep and the level of the river was so low he could have damaged it. We didn't really want to sit around all day waiting for the men to return, so we went to see the jungle on the other side of the river, and look at the work Victor's men had been doing.

The day's hunting produced nothing. Once, walking behind Katie, I saw her keeping her balance at a tricky part by holding on to a bough of a tree.

"Katie, take your hands off that tree," I snapped at her. I was very much for keeping my hands to myself, even tightly in my pockets, on walks like these. My brother had told me of the time his Indian guide struck his arm away from a tree trunk, at the same time pointing out a small black poisonous snake coiled at the very point where he had intended to rest his hand. I certainly didn't want Katie or anyone else to be bitten, and later on we were shown rather nasty inch-and-a-half-long biting ants that live on tree trunks, which seemed to underscore the moral of the story.

We inspected the deep holes, only a yard or so square though about four or five yards deep, which Victor's men were digging so that he could analyse the samples to see if gold was present, and if so how much. A mouse had fallen into one of these wells, and scampered round and round creating little sand avalanches — it buried itself alive. Working down these holes looked very dangerous to me, and I was glad to find out that some pulleys we'd brought, thinking they would come in useful for our gold operations, would instead be used for this rather different type of operation.

Though we fished for some time, our own lines looked ridiculously fine compared to the huge line I had borrowed in the early morning. We eventually used these thin lines to catch bait, small four-inch-long sardine-like fish which fitted the large hooks beautifully. But today just wasn't our lucky day, though it was for some of the others. As we returned to camp we saw the whole group of our men clustered round, and as we landed they held up a deer. If Katie and I had seen what else they'd brought, we wouldn't have even set foot on shore before they had got rid of it. But we didn't see, so we excitedly inspected the day's bag.

"Who got it? Where did you find it? How far into the forest did you go? Is it good to eat? How are we going to cook it?" Real venison, it sounded too good to be true. We had forgotten our pique at the men for slinking off without us earlier, though I noticed Victor taking the old man off for a few terse words. They had a baby tortoise, no larger than the size of the palm of my hand. They were going to keep it as a pet. Later on they presented it to Victor instead, together with a leg from the deer. But they then produced yet another "pet" they proposed to keep in camp.

Katie and I stared in horror.

"Get rid of it! Get it out of our sight!" Shivers ran down my spine. I scrambled up the bank to put as much distance between me and the new "pet" as possible. Coiled round his arm, Lucho had a large, live snake about six feet long, identified later as a mantona, a rather sluggish member of the boa family.

"Señorita Anna, it won't harm you. Look, it doesn't mind me holding it, it's quite friendly."

"No! I can't stand it! Get rid of it right away, I won't have it near the place."

If there is one thing I can't bear above everything else it's snakes. There are many other things I can put up with, but just seeing one like that gives me nightmares. I got my way. Katie photographed it before they threw it in the river, first fitting on to her camera the longest telephoto lens she possessed. We saw it swim away and it was only then that I could pull myself together, relax and enjoy the prospect of my birthday dinner.

The venison turned out quite delicious. A little bit tough, but then we hadn't a refrigerator here, and it would have been rotten in no time if we'd tried to hang it. The next day we had venison stew which had been cooked for some time and was even more delicious. The men had put some onions and salt and pepper, and favourite Peruvian herbs in it, and it was tender and seemed to be better than any stew I have ever tasted.

But we hadn't seen the last of the snake. It had swum off for less than a hundred yards. Then it must have come up the beach, and found some shelter in some low-lying bushes during the night. That morning Katie and I chose to film all day. We walked round and round, Katie looking for the best camera angles for some scenes in the movie we were making for the Banco Minero. Then she'd shoot while I stood by marking down notes. Suddenly one of the men pointed to a patch of greenery only a few feet from where we were busily working. There was our friend the snake again, looking a bit sheepish as if it knew we didn't want it around but hadn't been able to resist the pleasure of our company. This time Katie and I were so furious that the men scuttled around to catch the snake and take it away. One of the boys decided he wanted the skin so they put the snake in a sack and left it up at the camp. But we girls were to remain bad-tempered as long as that snake was in the vicinity.

"Why don't they just take it very far away and let it go?" complained Oliver. He was getting rather fed up with this female stupidity over something that was really quite harmless. Katie and I said no more, though later I began to feel rather guilty as the episode unfolded. The men must have changed their minds about keeping the skin. They opened the sack and did just what little boys in England might do to an injured bird. As it squirmed and squiggled they threw knives at it till it was pronounced dead, and finally threw it in the river. That was probably about the nearest I will ever come to being sorry for a snake and I still feel guilty about it.

That evening, two of the boys, Lucho and Alberto, who was vaguely related to Villanueva, decided they wanted to go and hunt a *tigrillo* — an ocelot. They had seen the tracks where it came to the river to drink every evening and were keen to investigate. The moonlight made walking along the beach such a pleasure that Katie and I clamoured to be allowed to go too. The boys took a gun, and I put on a second shirt so that the mosquitoes couldn't bite through, in case we had to sit still for long.

"Katie, are you going to wear your boots?" I called over to her.

"Well, I was wondering if we shouldn't wear our gym shoes. They'll be so much quieter on the beach."

This was our first moonlight escapade, and we wanted to be perfectly quiet so that we would be taken along on other hunting trips.

We went in a creeping single file. Every few yards Lucho or Alberto would stop and shine the torch around looking for animal spoor. We didn't talk at all and if they saw tracks they would whisper the name of the animal so that we knew what we might come up against.

Four of us was too great a number to go hunting. With the best will in the world we couldn't be as silent as one or two. But this evening was a practice run, and though we didn't see a *tigrillo* we didn't come home empty-handed either.

The single file suddenly stopped, and Lucho up front was ducking and signalling with his hands for us to freeze. He waited a few seconds, then charged forward at a sprint for about five yards and threw himself into a rugby tackle at something on the ground. He came back covered in sand, grinning from ear to ear and clutching a turtle.

"She was just looking for somewhere to lay her eggs, so I had a chance to creep up on her. But if she'd been able to run I would never have caught her," Lucho announced. I didn't know turtle was edible, I've only heard of tins of clear turtle soup which seemed to have no resemblance at all to actual meat.

"But it tastes really good, *señorita*. It needs a bit of time to prepare and cook, but if it's boiled for several hours it's good meat."

So we had plenty of food, though meat wouldn't keep twenty-four hours in that heat. We would have to eat all the deer we could manage tomorrow, before it went bad. Then it would be our turtle's turn for the cooking pot. That evening there was another example of the men's rather callous attitude towards animals which upset us all. When we returned, they put the turtle upside down into the large petrol drum where the poor creature scratched and scraped, desperately trying to right itself. Finally Oliver could stand it no longer, and told them to do something, because otherwise no one would sleep. They tied the unfortunate reptile to a post, but it escaped during the night. The three of us were secretly rather pleased.

We were unlucky in the night. A big wind came howling through the forest, and a spectacular storm brewed up, not far away. The sky lit up with forked lightning, and the rumbling thunder seemed to shake the very earth beneath us. It started to pour, and soon Oliver was complaining loudly.

"I'm getting soaked. The wind is coming straight at me, and if I stay here much longer even my sleeping bag will be dripping. I'm moving to the blue tent." The rain brought the temperature down dramatically, and I had to pull on a few layers of clothing inside my sleeping bag to keep my teeth from chattering. Katie slept through it all.

The rainstorm had blown over by morning, but it had left us

with a lovely cool day. It was wonderful. We had to walk around with sweaters on, and we felt almost as though we were at home. The insects had vanished, and everything was perfect.

Villanueva strolled over. He was a bit surly, and had hardly spoken to me since Sunday when he'd been told off by Victor over the business with the canoe.

"*Señorita*, the men won't work today, it is too cold for them. As they have to work with the water they would get sick. This is one of those summer cold spells—we get them every so often for a few days at this time of year. No one works."

So we had to accept it. It was very odd—we would have given anything for a couple of days off during that terrible heat. Today was just the right temperature for us to work, and now the men had downed tools.

CHAPTER SEVEN

Turtle-egg omelette

Burning the pans

As we weren't working I decided I would get the boys to take me turtle-egg hunting.

"On a beach opposite, Señorita Anna, there is a large strip of sand and a pool of water. It is just right for the turtles. They love a bit of sheltered beach on which to lay their eggs. It should be perfect."

So Lucho, Marino, Alfonso the *motorista* and myself set off to hunt more protein for our diet. The deer was finished, the turtle had escaped, and we'd been told many times that we had hit just the right season for *huevos de charapa*—turtle eggs. We wanted to find out if they were good to eat. Since our stocks of dried meat had dwindled far more rapidly than we'd expected, I was very keen to grab every opportunity of eating protein. There would be periods when it would not be available.

We secured the canoe by tying the rope round a heavy stone on the beach, and followed the edge of the small offshoot of the river. It was like a large pool, full of still, deep water, cut off from the main river by a sand-bar—till the next big rainstorm.

"What are those tracks, Alfonso?"

"Those are *largato*—alligator prints, *señorita*."

77

"They look very fresh to me," I said. The mud was all wet and even the claw marks showed up perfectly.

"We probably gave it a scare. It's run out of the water and the tracks go straight up into the trees," Lucho volunteered. And so we continued, clambering over tree trunks and squelching through heavy mud almost like quicksand. There was no sign of other tracks on the beach. We continued to struggle along till we came to a rise.

"This is where we must start looking carefully. The *charapas* only lay their eggs in the highest parts of the sandy beaches," Marino said to me. Everyone spread out, though I kept close behind Lucho to see what happened. He stooped down, pointing.

"You see those tracks? That's a turtle. She's walking in a straight line away from the river. We must follow."

I watched and followed Lucho as he continued, his eyes glued to the sand.

"Look, here she's starting to walk in a curve. She's looking for somewhere to lay her eggs, *señorita*. These tracks are two days old, and with last night's rain they are much more difficult to follow. But somewhere near here she will have laid her eggs."

The tracks looked so faint I could barely distinguish them from the tracks of birds and small animals. But they were different. Most animals cross their feet over and make an irregular trail. but these were two neat sets, perfectly parallel, nine inches apart,

"Here is where she laid them, this place where she has walked in a tight circle. The sand is all disturbed where she digs deep down and then buries the eggs, smoothing the sand over flat again as well as possible. You dig here while I look for another nest."

Gingerly at first, my hands scraped away the top layers of sand. But I looked up to see Alfonso some yards away digging with great vigour, so I put more zest into the operation. In one place the sand was all loose, and here I continued digging until I found the nest. It was over twelve inches deep and there was a cluster of little round white eggs, not yet completely hard. They were as big as a slightly undersize ping-pong ball. Getting them all out seemed to be a never-ending job as there were twenty-five of them. Finally, copying my companions who were leaving their little piles of eggs on the ground by the nests, I set out to look for another one myself. It was very easy. There were so many turtle tracks crossing one another that the evening they had laid their eggs must have been like a particularly busy night at a maternity hospital. I followed one set of very distinctive tracks of a turtle which had followed the sandy dunes before clambering up quite a steep rise. Then it had started its circular hunt for a good spot. I soon found the tell-tale patch of disturbed sand, and it was then only a question of digging.

"Marino, how are we going to get all these eggs back?" I called across.

"*Señorita*, I'm going to put the ones I collected in my shirt, but I have so many it will be quite full."

Then I took my hat off and it became a moderately efficient shopping bag. By this time we had plenty of eggs, for all of us had dug out two or three nests. We made our way back to the canoe and crossed the river to our camp.

"These eggs keep for over a week, and if we salt them they will keep for months, but one of the best dishes to eat when they are fresh is turtle omelette," Marino told me. "I'll make one for you right now."

I was pretty hungry, although it was only 7.30 a.m. We'd left without having breakfast and such an energetic start to the day had given me an appetite. Katie and Oliver had not eaten either; they were waiting for us to come back before preparing anything.

Marino took about thirty of the little soft eggs and broke them one by one into a plate. I saw him pour away a clear fluid that you get in these eggs. He then took a fork and sat there beating the eggs in a very relaxed manner for four or five minutes. In the end the stuff was like meringue mixture though very bright yellow, and not quite stiff enough to stand the fork up in it. Next he added salt, and poured the whole mixture in a hot frying pan to sizzle in cooking oil on his wood fire. He covered it with a lid and we waited. When he flipped it over the smell increased our hunger to dizzy heights. Then he pronounced it cooked.

It was like a spongy cake, well cooked and so delicious that we couldn't have eaten better in the world's best restaurant. It was food for the gods. I was pleased to see Oliver really enjoying his *charapa tortilla*. He wasn't liking the heat and the mosquitoes, but he would at least have pleasant memories of some of the unusual food we had eaten.

Now we had been introduced to this delicacy we were to have it many times, and with many variations. One day we were given some lemons, so instead of adding salt, Marino put a couple of spoonfuls of sugar in the mixture and we squeezed lemon juice over it. That was even better. Marino beamed. "I worked as a cook for a few months up in Cuzco," he said. It was a bit of luck this, because every so often he would bring us some new and tasty mouthful that we would never have expected so far away from civilisation.

I was in trouble with Katie later that day.

"Anna, I'm fed up with you burning the pans. I'll never get all this black off."

"Sorry, but I forgot to stir the soup again. It's these beastly insects," I added lamely. "When it's dark I can't see them settling on me, and I just get covered in bites if I sit out here over the stove. I should try and do the cooking before dark."

"Well, you can jolly well clean the pots yourself when you burn them next time," she groused.

In all fairness, it was happening with much too much regularity and I needed telling off many times before it would sink in that if I set the pot to boil I mustn't forget it so often. My other excuse was that the primus flame was difficult to regulate, and often the

Anna panning for gold
Banco Minero employee buying gold

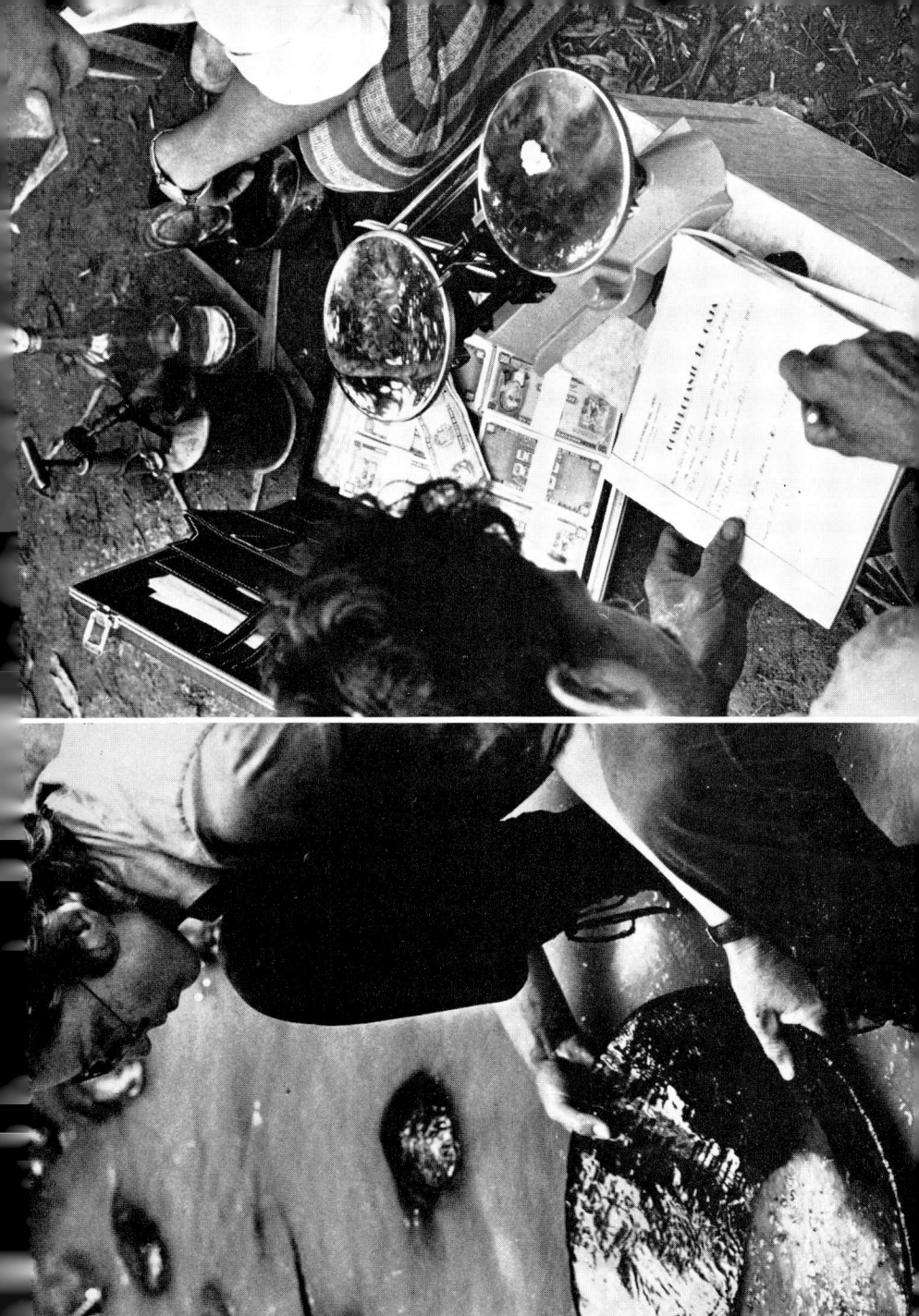

top looked innocently uncooked while the bottom was forming a congealed black mess.

"I discovered yesterday how the locals get rid of the grease and burnt spots on their pan," Katie told me, rightly anticipating that it would be several days before I learned to stop burning the pans. "I watched Horacio cleaning out that big pot the men have. He takes a handful of sand and grit from the beach and scours the pot with that. When he's got all the muck off, and the grease too, he finishes the job by rubbing a leaf all over it. It's a very good system." And this was how we washed our pans and dishes from then on: no detergent, and no hot water.

But Swiss Katie wouldn't pass any dishes that were not spotlessly clean. My London standards fell lamentably short of hers, so in spite of her griping about my burnt pots I was usually cook, and she did the washing up. Katie also kept us up to scratch with our tidiness. We had the big inflatable tent with our provisions, clothes and spare equipment stacked inside. Every item from food to clothing had its own plastic bag, and then this bundle went into a collective plastic bag with other bits and pieces. Both inner and outer bags were sealed.

I think I was the one who first said in exasperation: "People are getting up too late."

Katie picked it up: "People are too untidy. I refuse to clear this tent up again."

And Oliver chipped in with: "People are hanging their wet clothes up, and my dry clothes are getting wet from them."

It was a happy invention, that catchword. Whenever someone felt like bitching but didn't want to get too personal, out it would come, and in this way instead of having a bad atmosphere and arguments, everybody would own up to some bad deed, impossibly, and we would all start giggling and there was no ill feeling. Partly as a result of this verbal nicety we were able to follow through our original aim of remaining good friends.

Oliver and Katie were both having quite a rough time at the

View from Five Bend Camp
Oliver cutting through the jungle

beginning of our trip, which worried me because it happened so soon. How would they ever stick it for three or four months? Oliver had had rather exotic expectations of the jungle. He'd thought it would be much more colourful, full of brightly coloured orchids, brilliant birds of paradise, monkeys, alligators and all the other things you see in Tarzan movies. These things are all there, but the jungle hides its secrets behind a solid wall of green; the animals are wary of man, and tend to keep out of sight.

And Oliver also had to put up with the hard slog of physical work on those hot open beaches, which hit him hardest because, being a man, he did the heaviest work. He never complained or grumbled though.

The first part of our journey, with six awkward men to direct, was a real educational experience, looking back, but not altogether fun at the time. Oliver liked it even less than I. He had many constructive ideas about how to wash the gold or how to adapt the system in some small way to make it more efficient, but the men would do nothing new, were absolutely fixed in their ways and would not even listen. They always knew better, and this drove Oliver to despair several times.

Katie just could not take the heat at first and sat around miserably, looking pale and sickly. For her it went completely against the grain to be doing nothing, but the heat sapped her energy and natural high spirits. We were very lucky from this point of view that we had cool periods soon after arriving, so that there were times when she did a lot and enjoyed herself.

My bad patch was to come later on. After a while I was really fed up with our six man team, so wanted to take a smaller party to go much further upstream. But we'd had a message to say that my brother Nick and his son Igor were planning to come and visit us. As time went on and we had no further news I got very miserable and agitated. We would miss them if we took off before they arrived, yet there was no way of finding out when—and if—they would be arriving. I kept wondering what we ought to do.

Then one day when Katie and I were going through our diaries together a phrase on one of her pages caught my eye: "Anna is very touchy these days." After that I was able to pull myself together—shock treatment!

One of the things I enjoyed most about the whole trip was that at times we were living off the land almost entirely, and from this we had some of the most delicious meals we had ever eaten. There were usually fish to be had, different kinds, straight from the river and into the boiling-pot; our fresh turtle-egg omelettes were such a delicacy, and the wild turkey, cooked for us on a day when we were so hungry anything would have passed, was a meal I won't forget in a long time. Of course, too, it was often a case of feast or famine, that being the way of things with hunting. At times there would be so much to eat that we couldn't finish everything and wound up throwing rotten meat and fish into the river. Other times the men would come back empty-handed and it would be back to packet soups and dehydrated vegetables, which were so dull in comparison that we preferred to regard them as emergency rations.

We ourselves seldom did the hunting. I was a terrible shot, and couldn't even hit a standing tin can in practice. Oliver was a good shot, but he wasn't keen on the hunting trips. Katie was the champion of our trio—a fact she attributed to her precise, camera-trained eye. But our hunting wasn't for sport, it was for food, so the best shots—Lucho and Marino—took the guns.

Once the two boys announced that they would like to get *tigrillo* skins as presents for Katie and me. It was a kind thought, though it was difficult to explain that the last thing we wanted was two dead ocelots, just so that we could have the skins. If they could fix for us to film or photograph live ones—that, yes, would be a real treat. But this was to prove impossible, for the *tigrillo* prowls around only at night.

Having the two boys with us meant that we didn't have to

clean and skin the animals ourselves. This was a blessing, because, try as I might, I've never managed to clean even a chicken without feeling sick. Skinning and gutting a deer would have been quite beyond me. Once or twice I gingerly tried plucking the feathers from game fowl, but the boys were so much quicker and better at it that I could see no reason for not leaving it all to them.

The men killed a large tortoise on one occasion, and Katie wanted to keep the shell. We couldn't figure out how to get rid of all the flesh stuck to the inside, though, so we decided to consult Marino.

"We'll take it down to the far end of the beach and leave it out for the vultures. They'll scrape it clean with their beaks, and what they leave behind the ants will finish," he said. "We'll leave it there for a week or so, and then we can get rid of the bugs with insect spray." It was left there much longer, because we forgot all about it. Many weeks later, travelling down from a camp much further upstream, we saw it lying there on the bank. Katie now has it polished and hanging on her wall at home.

CHAPTER EIGHT

A prospector's prospector

A parking ticket

OUR FIRST COLD SPELL lasted for a couple of days, and the men continued to insist obstinately that it was too chilly for them to work. It *was* rather unpleasant at times, with intermittent rain and blustery winds, but the temperature was still pleasantly warm by our standards.

The day after the turtle-egg hunt I got Lucho and Alberto to take me for a walk in the jungle. I followed them over a carpet of soft, damp, rotting vegetation through an endless green-tinted gloom. Most of the time it was easy going here, and only occasionally did we have to use our machetes to chop through a dense spot. Sometimes there would be the trunk of a fallen tree to climb over—one of the great monsters that towered all around us. We could hear birds calling and screeching in the roof of the forest, more than a hundred feet above our heads, but we seldom caught sight of them.

The day was overcast, and there was no sun to give us our bearings, but I was nevertheless astounded when, after little more than an hour, Lucho confessed that we were lost.

"*Señorita*, I've only been lost in the jungle once before in my life," he said helplessly. He added, almost as an afterthought:

"That time I had to spend the night in the forest and wait until next afternoon for the sun to come out."

This was not calculated to cheer me up. How had we managed to lose ourselves, anyway? We'd been following a trail of sorts. The word "trail" needs qualifying, because what the locals term a trail is indistinguishable from any other piece of jungle to an outsider. Once during the walk out from the camp the boys left me to wait while they went off to chase a group of monkeys we'd heard chattering in the trees nearby. Maybe it was this hunting detour that confused them.

We walked around aimlessly for another hour. Lucho would frequently bend down to examine the ground for footprints, or he would call for us to be silent while he strained his ears to pick up some sound. I wasn't sure what he was expecting to hear. In spite of the present situation I still had confidence in Lucho. He was, after all, an Indian of sorts, probably with some Machiguenga blood, though he was from the relatively tame territory of the lower La Convencion, a long way down the valley from Cuzco. I wouldn't let myself believe that he was capable of getting us hopelessly and utterly lost.

"*Señorita*, I can hear the river. Listen," he said, after one of our ritual pauses.

I couldn't hear anything. He made a forty-five degree turn and motioned us to follow. Twenty minutes later I could hear the engine of some boat on the river, though I had no idea which direction it was coming from. It took us another half hour to reach the river bank, hacking our way through a patch of really thick undergrowth. We emerged half a mile upstream from where Alfonso was waiting with the canoe. We found him fast asleep, evidently unconcerned that we were hours overdue.

The sun stayed in again next day, but the weather was warming up again. The men decided they could start work again, so we set to, trying to make up for lost time. It was becoming very obvious to us that the first estimates of an ounce of gold a day from this beach just were not realistic. If the men made a habit of taking odd days off we were going to have trouble making ends meet.

The work, though, was disrupted once again, this time by the forces of nature. We'd just got the whole system set up, when Alberto, who had always been in charge of the motor pump, announced that the river level was rising. Sure enough, there was the pump standing in six inches of water, though chugging steadily on regardless, for the moment. The water pump was moved up the beach and we started again. But, as we'd been warned more than once, when the river starts rising it's usually in a hurry. The storms of the last few days must have been widespread, and now the river, very muddy and swelling as we watched, disturbed us all day. We moved our water pump continually from its clutches, but by late afternoon there was even some concern for our camp.

Villanueva came over and told us to pack everything up.

"I've been in situations where we've had to abandon equipment when the river flooded without warning," he said. "Today it has come up slowly, but tonight we will have to watch carefully."

We all went to get some sleep in our hammocks, fully-clothed, everything else packed away, and everyone ready to scramble for the canoe if necessary. Several times in the night I heard one of the men prowling around. I slept fitfully with the noise of the river waters lapping at the banks so near to our camp. Often I would get up to check the river level.

In the early hours of the morning after our canoe had floated to within a few yards of the bank where we had our camp, the river level steadied. When I'd seen it in the same place for two consecutive checks I relaxed a little and caught a few hours' sleep before dawn.

For the next days the river was dangerously high, but the sun reappeared — as did the mosquitoes — and we were able to carry on working.

There came to be a lot of unrest in the camp. Villanueva

wanted to move upstream but everyone else wanted to go back down to Maldonado for Peruvian Independence Day which was coming up in a couple of weeks' time. This was a big weekend for celebration, I was told, and most of the men wanted to leave. We weren't sorry about this, because they had become something of a burden. We were paying them £1 each per day and our gold production wasn't covering even their wages, let alone the extras like food and so on. But we couldn't be left with no one at all, and if we'd waited for someone to come up from Puerto Maldonado we would have wasted at least two weeks.

With this problem still troubling us, we decided to explore upstream, taking with us only Villanueva and Alfonso our boat driver.

After we'd settled the other four down to work one morning we set off upstream, taking with us shovels and pans to see if we couldn't find a good beach to continue operations after most of the men had left. It was refreshing to get on the river again. For some time we hadn't moved much, except to cross from bank to bank, keeping to a policy of travelling only when strictly necessary. The river was sometimes dangerous, with its hidden tree trunks, whirlpools and rapids, and this far upstream we had no way of replenishing fuel supplies: we wanted to be sure of having petrol when we really needed it.

Villanueva took up his stance at the bow of the canoe, constantly measuring the depth with his nine-foot bamboo pole. Half an hour after we'd started, we came up to a place with a small red and white Peruvian flag flying high in the breeze at the top of a very home-made looking flag pole—it was more like a dead tree, propped up. After a signal from Villanueva, Alfonso pulled in.

"This is a shop," Villanueva announced. "A couple of the boys have asked me to get them cigarettes." We were really amazed: a shop in the middle of nowhere. Whoever were the customers? We climbed up the bank to the accompaniment of loud barking from three mangy-looking dogs. Katie was very nervous, since she'd been warned repeatedly before coming that South American dogs bite without reason, were vicious, and were all infected with rabies. I myself had bad, though less dramatic, memories of Peruvian dogs. A couple of times I had caught fleas from them. These ones were noisy rather than dangerous and we followed the pathway, letting the men go first. After passing a kind of vegetable patch with no vegetables we came to a group of huts with chickens, barefoot children, cats and more dogs running round joining in the excitement of welcoming the new visitors. They couldn't have had many customers, judging from this reception. Everyone came in from the clearings round about to have a look at this novelty — customers! There were some workers with machetes in their hands and women with babies slung across their backs in a blanket. Then came a taller, fatter man who definitely looked like The Boss.

"*Buenos dias, señores, señoras.* Please come over to my shop." And he led us over to one of the further huts, all of which were raised on stilts about three feet off the ground. We felt rather embarrassed at being received like honoured and wealthy clients, since none of us had even a *centavo* in our pockets. Money had been far from our thoughts this morning, and it had not occurred to the men to mention the local shop though we'd been living so near to it all this time.

We needn't have worried. There was almost nothing to buy. The dusty shelves held half a dozen tins of sardines, a few more of tuna fish, a couple of bottles of Milk of Magnesia, two crates of Coca-Cola and Fanta, a few rolls of toilet paper, some rusty tins of tomato sauce, tins of sausages and a few packets of cigarettes.

Villanueva bought some cigarettes and said afterwards that they were very expensive, three times the price they would pay in Puerto Maldonado. The shop keeper invited us to drink a Fanta, which we gratefully accepted and then we were on our way.

"That was Señor Orosco. He is well known in this district. He has more than fifteen men working gold for him on this river and down on the Madre de Dios," said Villanueva.

"But he looked so poor, he had no shoes on and his clothes were tattered," exclaimed Oliver. Villanueva shook his head.

"Señor Orosco owns a house in Cuzco, and has a lot of farm-land on this river. He has lemon trees and banana trees and he often takes his family down to Puerto Maldonado, or to Cuzco where his boy is at school."

It was hard to believe that anyone with any money at all could live in such apparent poverty. The open huts had rough beds made from bamboo. We didn't see mattresses, only blankets. It was difficult to credit that the man was a prosperous goldminer, but we were later told by the Banco Minero that his gold production was usually between one and one-and-a-half kilos per month, worth roughly £1,000 to £1,500. He wouldn't have been paying taxes on this money, for Peru has a blanket tax-exemption for all individuals and companies working in the jungle—a measure intended to encourage development of the region. So even after paying fifteen workers their wages of 60p. per day old Orosco was making a handsome profit. And this was not counting the gold he was certainly hoarding.

Our journey continued. We passed one more settlement on a bank high above us where a line of small, laughing faces pursued us, bobbing along behind the bushes. The children called and waved until we slid out of sight behind the next bend.

The river was so shallow in places that we had almost to stop the motor several times. We picked our way through a tangle of fallen tree trunks, thrown up by the river floods, that became worse the higher we travelled upstream. In one place there was a group of men working for gold on one of the beaches.

"Those are Orosco's men," Villanueva called out to us. These men were working in much the same way that we weer, but they had a bigger trough. Instead of working from the side as we did, they had a man standing on top of the trough shovelling away the washed material. We waved and rode on.

Finally Villanueva signalled for Alfonso to pull in to the side, at the upper end of a long beach covered in driftwood. Our foreman said that this was a typical position and angle of the river that might give us a good gold-producing location. Oliver and Alfonso started working on the tryouts while Katie and I inspected the beach.

It was here that we started our impractical collection of pretty stones. We were first attracted by the astonishing variety of colours and patterns in the stones on these beaches while we'd been doing our stints as "washer" back at the sluice trough. Then we'd been far too busy to stop when we saw a nice one, but now as we walked around the beach we began to pick up fancy stones that caught our eye. Katie searched for green ones, which her mother particularly liked. I was less selective and scooped up anything that specially appealed to me, and wasn't too big. We had become accustomed to travelling with plastic bags in our pockets for collecting the samples from our tryouts. I didn't believe in wasting any gold as quite a lot of effort went into these tests. From today onwards, though, there were extra bags for coloured pebbles.

Our stone collection became a hobby for the rest of our trip, to the despair of Oliver, and anyone else who had to help when we moved camp. We ended up with a huge bag of stones, enormously heavy, but alas it was all for nothing. On the final canoe journey down to Puerto Maldonado they disappeared. Someone had either taken them to be worthless and thrown them overboard, or assumed they were valuable and stolen them.

"How's it going, Olly?" Oliver was sweating over his third panful. We were all anxious to find somewhere really good.

"Not so hot, I've done three samples, all from different places, and it isn't even up to our own beach."

Alfonso shook his head. He hadn't had **any** better luck.

"Let's move on then. There's a big beach higher up."

We moved a little way upstream, and the men started taking more samples. I sat in the canoe munching 'miner's biscuits' — ghastly things that form a main staple of the working men in this region. They tasted like baked flour-and-water, but we had some Marmite to jolly them up a little. I lay back to relax for a while, and soon fell asleep in the hot sun. When I woke the river had risen. The canoe was firmly tied to a tree trunk, but it was now several yards from the shore — just when I'd been hoping to spend a whole day with dry feet. I waded ashore, grumbling to myself.

It was time I took an interest in our future gold production. There were little holes spread out over the beach area but the beach was now deserted. As I neared the tree line I could hear talking and digging, and there a few yards into the jungle was Villanueva standing in a hole four feet deep trying to see if there was a good gold bearing level well below the surface.

"This would be a good patch to work, but we'd need twenty men to take away the top soil, so we'll have to give up on this one."

Villanueva pointed to the opposite bank.

"Someone lives over there. Look — banana trees. And there are a couple of canoes. Shall we take a look, *señorita*?"

I nodded, and we crossed. The river was bringing down so much débris that Villanueva had to fend off logs with his pole. As we drew in there was the usual feigned assault by a group of scruffy dogs, barking and snapping. Then a dark, youngish man with sharp, bright eyes and a beard and quite long, black hair — obviously an eccentric, this one — came striding forward with outstretched hand and a big smile on his face.

"Welcome. Come on up, see the house and visit my *chacra*." We were lucky to find him here, he said. "I'm usually away on the river, but today I'm working with my men clearing away some jungle to make a new plot.

"We don't get many visitors here," he added. "But I was expecting you people to turn up sooner or later."

We were to find wherever we went that the news of the three gringos panning gold had got there well ahead of us.

We all clambered up the bank and followed the man through an avenue of banana trees. Everything looked neat and well tended. We had seen other attempts at farming, but the plots were usually untidy and overgrown. Here the grass between the plants was cut short, and there were even rows of *yuca* — a starchy root vegetable — growing along the edges. Chickens and pigs scattered as we came to a couple of huts in a clearing.

"Come and sit down and have some coffee."

Our new friend ushered us up the steps of an open-sided hut and motioned us to sit at a bench. Clapping his hands he directed one of his men to put the kettle on a fire in the yard, which was obviously kept on the go continually. Some other big pot was steaming gently on the fire, balanced on a metal tray that I guessed had once seen service as the stone-filter on a gold sluice. Fresh wood was stuffed under the tray, and a man knelt down, blowing with all his might, puffing up a cloud of ashes, and fanning furiously with a saucepan lid to kindle the flames.

An old mother bitch lay sprawled near the hut, panting in the heat. Her offspring formed a struggling mass, pushing and shoving around her stomach, and tiny chicks swarmed, hopping over the puppies, picking at their coats, looking perhaps for fleas. In the hut we sat among piles of dried maize scattered over the dusty floor, on a platform festooned with faded clothing, hammocks and mosquito netting, all hanging from the rafters.

"Would you like to eat some *choclo* later on?" our host enquired. *Choclo* is corn on the cob. It sounded like a good idea to us, and we said so.

"Good. I'll get one of the men to pick some fresh ones. Here, let me show you my *chacra* while it's being prepared."

We strolled after him through long grass, stepping over tree trunks and halting every now and then while different crops and plants were pointed out to us. There was a big patch of sugar cane, lots of papaya trees — like a palm, with a tall smooth trunk — coffee plants, a lemon tree, an avocado pear tree, maize and *yuca*, shrubs of the violently hot pepper called *aji*, and even tomato plants, loaded with tiny fruit, the size of a marble.

"Just now I have six men here," said our friend. "I really need fifteen. What I want is to have half the boys working for gold in the river, and the rest working here on the farm. At the moment we do a few days gold and a few days farming each week.

"By the way," he added, with a smile, "my name is Wilfredo Segales."

Our guided tour ended with the news that the kettle was now boiling. As we made our way back to the hut I explained how it was that we three had come to be in the Madre de Dios. He showed polite and, I think, genuine interest in our plans and seemed more willing to take us seriously than most of the people we'd met so far.

"Listen, if you want any help then just ask me," he said as we sat drinking delicious coffee. "I can show you how we work for gold and I can take you hunting if you're interested."

This was the first really enterprising person we'd met living on the river. All the others there seemed to be eking out an existence, living day to day, never looking beyond the margins of a crude and uncomfortable life. Wilfredo said he wouldn't be staying in the jungle for ever — he wanted to build up enough capital to start up a small bakery business in Cuzco or Puno.

"Do you live here all the year round?" Even surly old Villanueva was interested to find someone like this so far from anywhere. "I mean, do you ever go into town?"

"Oh yes," said Wilfredo. "My men never stay for more than three or four months, so I have to go to Cuzco now and then to hire workers. Then I have a good time, have a little fun, take out girls, see a couple of films, and have a few beers.

"I have to buy stores when I'm up there, too. I'm usually away for about a month in all. There's just one person here to look after the animals."

Villanueva was curious, and kept probing: "Why haven't you got a woman here?"

Wilfredo sighed. "Well, I've asked plenty to come, but this life is too hard for a woman. I don't believe in lying too much to them, and once they've heard about the insects and the heat, and that there aren't any shops or electricity — well, they just don't want to know. I suppose I'll have to go and live in town when I get married. I won't be putting it off for many more years, because I'm already forty-one and girls don't like old men."

"Oh, you can't be forty-one!" I exclaimed. "You don't even look thirty."

The three of us had already speculated about his age, because he seemed so young to be owning a fair-sized farm and bossing a group of men. Wilfredo laughed at my reaction, but he evidently wasn't joking about his age. Later, when we had known him for some time, we learnt that his youthful looks and vigour concealed an impressive maturity and depth of knowledge of the jungle and its ways.

A large bowlful of steaming hot corn on the cob was placed before us. It was delicious, so tender and sweet that even Katie with her sparrow's appetite polished off three with no hesitation at all.

Our plans to continue upstream had been very pleasantly interrupted, and now, after taking a look at the river, we decided to abandon them altogether for today. The level was up again, and the current was running more strongly than before. We got up to leave, and Wilfredo loaded us with gifts of bananas, papaya and yuca. We promised to return soon, when the river was down to its normal level.

Luckily for us Alfonso was a skilled and prudent driver, with none of the nonchalant bravado that seemed to characterise most of the other *motoristas* we had seen on the river. On the way downstream we jockeyed for a safe channel among great rafts of swirling débris, with huge, partly submerged tree trunks dipping and heaving all around us. Now the turbulent river was so red in colour—that's presumably how it got its name—that it was hard to imagine how our filter would ever get the water clean enough to use for cooking. The storms that were responsible for all this finally arrived in our area about twenty minutes before we reached camp. Inky clouds blotted out the sun, and the rain pelted down so suddenly that we had to scramble to cover our cameras. Back at the camp we spent a desperate few minutes rushing around putting all the loose items of gear under cover.

Soon after we arrived we got the nicest surprise anyone could have hoped for. A canoe came sliding unnoticed into the bank, and suddenly there was Victor Vargas, grinning and waving a fat package. The post! Letters from the outside world—our first news in weeks. Katie disappeared with a large pile. She must have had a letter for every day we'd been away, so we saw no more of her that day. Oliver and I chatted with Victor, who wasn't in quite such a rush as usual.

"Anna, we've had a radio message from Hugo Fuentes. He's coming upriver to visit you, and he may arrive tomorrow or, if not, the following day," Victor told me. "So many people always want to see him that we're never quite sure when he'll arrive, though."

Hugo Fuentes was, we discovered, much in demand on the river. He was in charge of the whole of the gold operation that the Banco Minero was just setting up in the Madre de Dios region, covering several thousand square miles of territory. The Madre de Dios includes many large and hundreds of small tributaries of the river itself so Hugo Fuentes spent a lot of time travelling. The price of gold had gone shooting up from $64·00

*We feed a dozen **Indians**, Five Bend Camp*
Sunday football at Wilfredo's place

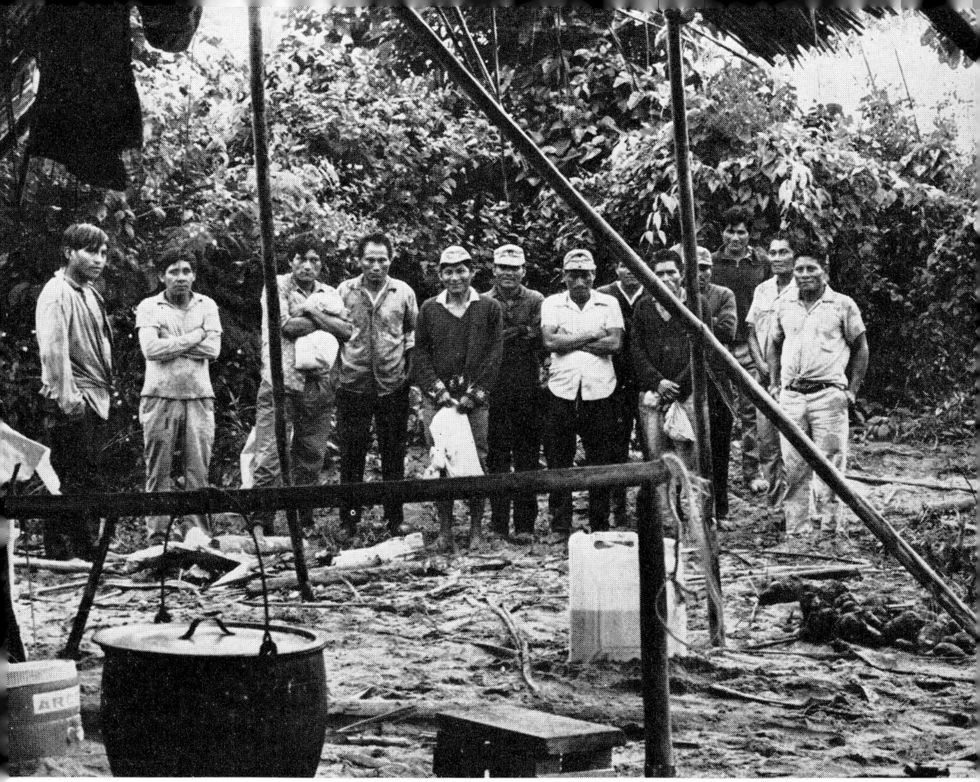

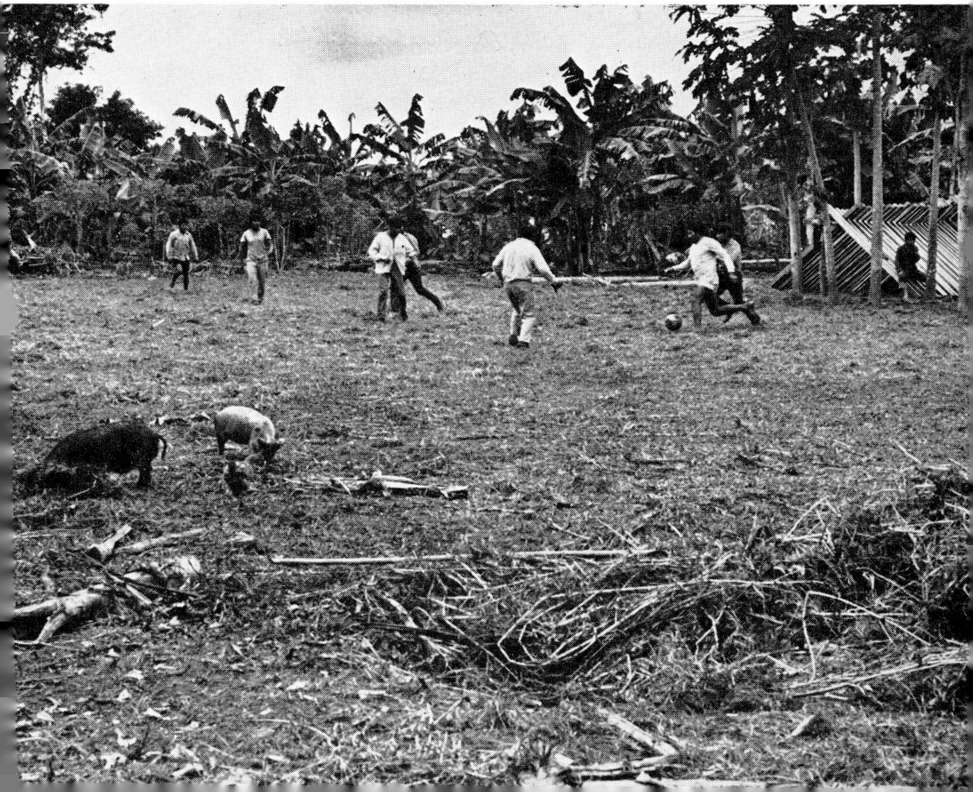

an ounce in June 1972 to $127·00 an ounce in July 1973, and the Bank had been designated by the government as sole buyer of any gold produced in Peru. Its prices kept pretty much in line with price levels on the world market, though there were some deductions for refining, and they paid with local currency, not dollars. Banco Minero were encouraging as many people as possible to come and look for gold, and anyone coming in was being helped tremendously with grub-stake loans of supplies and equipment to help start them off.

We had just arrived as this project was getting off the ground, and Hugo Fuentes was delighted to see international interest in their work at such an early stage. We found that Hugo's authority and his opinion and judgment were highly respected by all the miners, and they would save up any little problem they had until one of his periodic visits. Everyone would know he was coming. Most likely they had heard on the news from Puerto Maldonado over their transistors. The upstream progress of the Ingeniero's canoe would frequently be interrupted by some small figure on the bank, shouting and vigorously waving a shirt to attract attention.

Oliver was startled to hear me utter a loud groan as I opened my mail. I'd been far too efficient in getting a friend to forward *all* my letters. There were two of those impersonal buff envelopes that spell officialdom; one contained my telephone bill, and from the other fell a slender sheaf of parking tickets from the West-minster City Council. I got some pleasure from putting them on the camp fire. They would be out of date in any case by the time I returned.

What with the bad weather and our day off to go upstream looking for new locations, our gold production was suffering considerably. We decided to put in a few days' really hard work. The weather turned cool again after this latest storm, but not so cool that the men wouldn't work. Whoever did the washer or stone turner's job was subjected to something like a cold shower for four hours at a stretch, so Katie and I decided to make our anoraks available for anyone doing these two tasks. These light, knee-length garments were absolutely waterproof, so it was possible for anyone who was careful not to get the stream of water anywhere near his neck to walk away after a morning's work with only his feet wet.

It was very hard to understand this weather. We'd been told this was the summer season, but for short periods it was quite cold. Certainly we had never expected to use our sweaters, but it was as well that we'd brought them, for at times they were in-dispensable. Someone explained that winter was hotter than summer in the jungle, and it rained heavily for long periods — (which, in our limited experience, was when it turned cool!). Summer was cooler, and it rained far less frequently. It was all very confusing and we came to the conclusion that the terms summer and winter as we understand them had no real meaning in this part of the world.

No one in Lima had told us that it might turn chilly; we'd only been warned not to extend our stay into the rainy season, or else we'd be stuck there until the following summer. This state-ment wasn't true either, though again we had to go there to find out. There *was* a way out, not through Puerto Maldonado, but up the Rio Inambari and by truck along the road to Cuzco, following the route that people like Wilfredo Segales used all the year round, but this road could take two weeks if the rains were really bad. Puerto Maldonado airport was open every now and then in the winter too, it just depended on the weather.

One evening, when the river rose threateningly yet again, our neighbours, the two men we felt had jumped our claim on the nearby path of dried river bed, came to visit us with a little boy. It was nearly dusk, and I thought, since they just stood around chatting to us for some time, that they'd come by to see how our camp was progressing. At length, as darkness was closing in, one of the visitors suddenly blurted out:

"*Señorita*, we're in a bit of trouble. All our equipment is down on the bank on the other side of the river, and the river is coming up so fast I'm afraid we might lose it. Could you lend us your canoe?"

"But of course. Hurry though, look how dark it is."

I called over to our *motorista*: "Quick, Alfonso! Get the canoe and take these people across before it gets too dark."

Their little boy solemnly went with them. He was only five years old, but he had no other children to play with, so he copied the men and behaved just like them. Probably he wouldn't have had the faintest idea what to do with a toy; he was there to join in the work, and that was his life. It was sad and rather pathetic to see a small child so utterly robbed of his childhood. There was never any sign or talk of his mother.

The men rescued their equipment, putting it high up the beach, and next day we saw them crossing the river on a makeshift raft they had built. We were often to notice how withdrawn and diffident were these humble labourers and peasants of the Peruvian jungle. They would never push themselves forward, always accepted their poor lot without question or protest. Only in dire circumstances would they assert themselves as those two men had done. Even then they had stayed talking to us for a valuable half an hour of light until it was nearly too late before summoning up the courage to ask us this simple favour that they so urgently needed.

Two days after Victor's visit Hugo Fuentes arrived, smiling and very jolly. He'd brought us some beer, a hat for Oliver and some shotgun ammunition, which we were desperately short of. Marino was by now a marked man as the talented chef of the party, so I waved my arms around grandly, directing him to make a turtle-egg omelette to tempt our visitor. We cracked open a few bottles of beer and had quite a feast, which we rounded off with freshly stewed apples. Well, dried apples boiled up with a little sugar, actually, but it tastes just the same. Then we had a hurried letter-writing session. Hugo was on a sort of whistlestop tour of the Madre de Dios and Colorado rivers, sorting out the Bank problems at Camp Colorado that had accrued since his previous visit, and he was expected back at Puerto Maldonado next day. Those particular letters took only twenty-two days to get to London. Others we sent later on were to take six weeks.

It was time for us to decide what to do about our men. We didn't want to keep them all—we couldn't afford it in any case. But we wanted Lucho and Marino to stay with us. They were always cheerful, those two, laughing and singing around the camp, and they were the only ones who liked to go off hunting and fishing, often very successfully. Alfonso we would miss. He quietly got on with whatever had to be done, he was often smiling and gave us the solid feeling that he wouldn't let anything bad befall us—he was the oldest of the group with Horacio, and they would both be a loss to us. Horacio we'd misjudged badly at first. He had the face of a real villain, and we thought he was going to be number one troublemaker. Instead he turned out to be the hardest worker. His face was deeply lined, he somehow managed to have a continual two-day-old beard which made him look rather frightening but after a few days we didn't notice it any more. He wanted to go back, though, with Alfonso for the Independence Day celebrations.

One problem was that the two young boys refused to stay if Villanueva stayed. They reckoned he didn't spread the jobs out fairly, but Villanueva himself hadn't said anything about leaving, and neither had his stepson Alberto. Katie and Oliver had got on well enough with Villanueva. Several evenings Oliver and the old man had played chess together, and they discovered a mutual interest in sailing. Villanueva was fascinated by the sea—though he had never actually seen it—and had read dozens of adventure stories about sailing. He questioned Oliver closely about the nautical complications of the Ra and Kon Tiki voyages, while Oliver listened intently, trying to understand everything

that was said. Before Villanueva left us Oliver made him a
gift of his watch, a gesture which pleased the old man so much
that he was positively affable as we parted.

But for my part I never got on well with our foreman, and as
the weeks passed this had developed into a fairly hearty dislike.
Partly this was because as boss of the party it was chiefly my worry
that he wasn't making any effort at all to boost our gold production. We had hired him for his experience and knowledge, yet
in the event he was worth no more or less to us than any other
labourer, because he didn't seem prepared to put his talents to
work on our behalf. His attitude was always: "Whatever you
say, *señorita*."

Once, for instance, we were testing a beach, and I asked,
"Should we work here?" his exact reply was: "*Señorita*, if
you think this is a good place to work, yes we can work
here."

Another time I wanted to go fishing, and I asked if it was a
promising day.

"If you want to go fishing, *señorita*," he said, "then of course
we can go."

This yes-man attitude was hardly calculated to inspire a good
working relationship. If he had been a jovial character, full of
good gold-prospecting and hunting anecdotes, it might have
been tolerable. As he was dour and taciturn I ended up counting
the days until we could get rid of him.

I hated dealing with these problems, though, and when it came
to the crunch of firing Villanueva, I ducked out and left it all to
Victor. Ultimately we just kept Marino and Lucho.

Alberto had loyally stood by his stepfather and elected to leave –
he'd been in an awkward position because he had wanted to join
in with Marino and Lucho but repeatedly got a cold shoulder.
He wasn't a good hunter or a good worker – he tried his best but
didn't really fit in.

With Peruvian Independence Day only a few days away, the

four men left us early on the morning we were preparing to move upstream.

After handshakes all round and promises to look them up in Maldonado on our return they climbed into the orange Banco Minero canoe and soon disappeared round the first bend. We were too busy to think a lot about their departure, though we were sad to see Alfonso and Horacio go.

We were moving upstream to the beach that had showed reasonable results on our earlier visit. This would give us a new section to work for gold and a new piece of jungle in which to hunt.

CHAPTER NINE

Five Bend Camp

Bamboo beds

VICTOR VARGAS TURNED UP at our camp to find us putting the finishing touches to our packing. The camp had that abandoned look, with nothing left but the bamboo skeletons of our shelters.

"I'm going down to Puerto Maldonado tomorrow for two weeks' leave," he said. "You'd better borrow my radio, then if we have any messages for you we can put them out on the Banco Minero's programme on Radio Madre de Dios, between seven and seven-thirty in the evening. Besides, you need a bit of entertainment out here in the wilds."

Throughout our three weeks with the men they had moaned regularly because we had no transistor radio. It was like depriving an English workman of his tea-break. We had never thought of bringing one along with us, and later, when someone in Maldonado had offered to lend us one I'd refused it. Peruvians have a habit of leaving the things on full blast round the clock, and though I'm generally easy-going about noise, I felt that out there in the middle of the jungle we should have a chance to listen to the sounds of the wilderness — the birds, the crickets, the river — and leave behind for a while the electronic noise of civilisation. So I dug my heels in firmly: No radio.

But now we'd reached the stage where we were always longing for any snippet of news from the outside world, and we were being offered a chance to receive direct information via the radio: when we could hope for another batch of letters: when my brother and nephew would be arriving; how soon we could expect a replacement for Alfonso the *motorista*. We accepted gratefully.

As it turned out, the radio was to introduce a new element into our lives: the nightly game of ploughing through waves of static and interference, trying to find Radio Madre de Dios. It never seemed to be in the same place on the dial twice running. Every morning we'd sit hunched round the little transistor, sometimes straining to catch the words through a loud oscillating whine, sometimes listening comfortably with a good reception. We had met the radio announcer in Puerto Maldonado, so he often threw in a few words of greeting for us on his programme, and would usually say a few personal and rather flattering words when he had a special message for us. The fellow had a very arresting voice and a way of rolling his Rs lavishly that made it very easy to know when we'd hit the programme.

All the local stations—there seemed to be at least one for each town—carried dozens of these personal messages on every programme.

Typically, the Banco Minero might send out a message "Miner Diaz—it has been reported you are not allowing other miners to work on the beaches near to you—please do not molest any person coming there—they have every right to work where they want," or a Señora Manuela Chavez in Cuzco might send a message to her son Jaime, panning gold somewhere on the Rio Inambari, that his father was very ill, and could he get home as soon as possible.

Since there was no telephone system, and the mail was so slow and unreliable as to be virtually useless, the transistor radio represented a giant stride in communications for people in the Madre de Dios.

The station we most wanted to hear — Radio Madre de Dios —
was the one with the worst reception. Later we were told that the
Puerto Maldonado station's generators weren't powerful enough,
and we were lucky to receive anything at all. Once or twice
Oliver picked up the B.B.C. World Service, which came over
the air as clear as a bell: in this way we would learn that Australia
had been all out for 239, that the wretched French were about to
let off another bomb in the Pacific or that thirty-two people
had died in a train crash in Greece.

Our camp move was to take us an hour and a half by canoe
further upstream. We had finally wound up with two full canoeloads of gear. It sounds a lot for five people, but it included our
water pump, wheelbarrows and all the rest of the gold working
equipment, plus the remainder of the provisions that the men
had brought, and which were intended to tide us over for the
rest of our stay.

It was when I had inspected these supplies that I realised what
an awful mistake I'd made in leaving the men to choose their
own menus and eat what they pleased. Naturally enough they
had steadily eaten their way through all the interesting things like
fruit salad, condensed milk, sugar, meat sauce, Nescafé, and cocoa.
We ourselves had hardly touched these stocks, believing it
better to save them until we'd finished our supplies from England
— which were also, of course, the more appealing items. That
left us facing the awful prospect of several weeks subsistence on
rice, dried beans, *fariña* — a tough starchy, Grape-Nuts-like flour
made from *yuca* — and porridge oats. There were only some
herbs, a few assorted tins of tuna fish, and a handful of sugar to
brighten up this dismal fare. No doubt about it, we were going
to have to do a lot of hunting.

Victor Vargas had made up his mind to accompany us upriver
to check that we laid out our camp in a good spot. Lucho and
Marino were good boys, but they were both barely in their
twenties. And Victor had decided somewhere along the line

that he was responsible for our safety and well-being. We never really understood why, but it was a comforting feeling and we were all grateful to him. Besides the radio he lent us his collapsible camping table, and a pressure lantern that worked on the same principle as a primus stove, and provided a really brilliant light. He also volunteered to do some shopping for us in Maldonado. The list included soap, shampoo, insect repellent spray, chocolate, an Indian bead bracelet (a present for Katie's birthday, which was coming up soon) a chocolate birthday cake, and three wooden golding pans. Victor blinked a bit when he saw that lot, but he said he'd see what he could do.

One item on that list we were particularly anxious for was the gold pans. Now the men had left us we only had one between the five of us. We'd tried to get them made locally, without success, and we'd sent many desperate pleas for new ones to Maldonado, with the same negative result.

Our journey upriver was very slow. Both the boats were very heavily laden, and though we were lucky with the river in the sense that it was still high, and therefore easier to negotiate, the current against us was very strong. We arrived at our new site mid-morning, and we had still to send our canoe back downstream for the second load.

Meanwhile we all paced around our new beach looking for a suitable campsite. It had to be as high above the river as possible, and we eventually settled on a sandy spot with a small clearing. Katie was muttering an old homily that goes roughly "Only a fool builds his house on sand", but we countered by suggesting that it would take an even bigger fool to try hacking a clearing out of the dense jungle further in from the bank. We set to work with machetes, clearing away the low vegetation, and soon we had enough space for the camp. I was just mopping the sweat from my brow after digging two deep holes for my hammock poles, when Lucho approached.

"*Señorita*, maybe you don't need the hammock. Marino and I

were thinking it would make a much better camp if we built beds of bamboo for you and the Señorita Katie."

"But, Lucho, I hate the idea of ants," I said.

"That's no problem—you'll see. The beds will be high off the ground. Okay?"

"Okay, Lucho," I said. Okay had become hip slang among the locals since we'd arrived. It was by now fashionable to know a few words of English on the Rio Colorado, and soon wherever we went we were being greeted with "Goodbye" and told something was quite impossible with the words "All right".

The boys went off to cut bamboo for various shelter frames, and for our beds. Victor did a sample panning of the gravel from my redundant hammock-post holes to see if we were building our camp on gold. We were, as it turned out, but there wasn't enough to justify working there: down on the beach was better. Katie got busy sorting out the film, finding a patch of shade for it, and organising the cameras. Thanks to Katie's diligence in keeping an eye on our photographic equipment our cameras all survived without mishap, and no films were ever left out beneath the blazing sun to overheat and spoil.

Oliver set to cutting wood for the fire, and I decided to get a meal going. By the time the second canoe had been unloaded everyone would be very hungry. As the last items were unloaded someone said: "Where are the shotguns?"

We all looked blankly at one another, and our hearts sank. Where *were* the shotguns? After talking to Lucho and Marino we discovered that the two guns we'd been using in the lower camp had actually belonged to Villanueva and Alberto. This was a catastrophe. I'd always believed that the guns had been lent to us by the Banco Minero along with the rest of our equipment. Our diet promised to be fairly tedious even with the fruits of our hunting. And now we weren't even able to hunt.

Victor came to the rescue. We must have a gun, he said. It was impossible for us to be without one. He made off for the canoe,

waving his hand reassuringly. Not to worry, he would go down
to Camp Colorado and be back in a few hours with a gun for us.

When he returned he had brought us a gun. But, he said
apologetically, he had only been able to get ten rounds of
ammunition. (The other ammunition Hugo Fuentes had brought
us had, we assumed, been taken by the men.) This was to cut down
our hunting somewhat, and to make matters worse, Marino
later went out several times and returned empty-handed — much
to Lucho's disgust — having fired and missed. We were to have a
rough time for three weeks until we were able to get some fresh
ammunition.

At our new camp (we called it Five Bend Camp, though
some time after we'd left the area I heard that the place was now
called Playa Anna — Anna's beach) we decided to abandon our
primuses, and really do the cooking local style with a wood
fire. Now we only had two men we'd decided to do all our
cooking and eating together. We had four large, smoke-blackened
witches' cauldrons that the men had been using, and they somehow wouldn't have looked right on a primus. Instead they
bubbled merrily in the flames, hanging from a horizontal bamboo
bar, which was itself supported at either end by two upright
bamboo poles. The fire gave a homely atmosphere to the camp;
it was quieter than a primus which makes a loud roaring noise.
It was also somewhat less practical. If the wind kept changing
direction the cook would be forever dancing around the pots,
screwing up his eyes to avoid the worst of the smoke and shedding
tears copiously into the stew. Sometimes the wood was too damp
or the wrong kind, and one would have to keep fanning away at
the faltering flames with a saucepan lid. This was all elementary
boy scouting, I suppose, but we enjoyed it anyway.

Victor set off in the late afternoon, after having devoted his
entire day to helping us move camp. He took some letters to send
to Lima for us, and said he would be back in about two weeks,
bringing as many of the items on our shopping list as he could

muster. He departed with a word of caution to us to restrict our travelling on the river as much as possible, now that we were without a *motorista*. Both Oliver and Marino had been given instructions about starting the motor and handling the boat, but it was generally acknowledged by everyone we spoke to, that driving a boat on these rivers was not something to be treated casually. There were tricky currents, sandbars, unseen shelves and sunken logs. In many places there was only one passable channel out of several apparent alternatives, and to pick the wrong one could be disastrous. We had no great plans to do much river travelling until we had received a replacement for Alfonso.

As it turned out we weren't able to do any travelling at all. For the whole two weeks we were without a *motorista*, Oliver and Marino were unable to start the motor. Both of them spent hours checking the spark plug and fiddling with the carburettor, but they couldn't get a peep out of it. Eventually we discovered that the two-stroke oil/petrol mixture had separated out because the storage drum down at Camp Colorado had been standing around for so long, and we were trying to run the motor on almost pure oil. Looking back, I think probably down at Camp Colorado they may have known about this but thought they would keep quiet about it. A sort of insurance policy to protect us.

The atmosphere was really relaxed and gay as we settled down on our first evening at the new camp. It was our first evening with a radio, and it was rather pleasant sitting by the fire listening in while we had our first supper with just the five of us. We were all so much happier here, the camp was better laid out, and because of our amiable and easy-going relationship with the two boys we were a group together instead of being Them and Us.

We went off to bed early. Now I had to come to terms with a hard new bed of bamboo slats. I realised as soon as I lay down that I couldn't possibly sleep on it as it was, so I got hold of one of the thick plastic awnings that wasn't in use, and folded it about eight times, making a layer at least an inch thick. Then I laid out my sleeping bag and crawled in. No good: I wasn't going to sleep. The hard, corrugated surface dug into my shoulder blades and my backbone, and every other salient part of me. Maybe I was being like the princess with a pea under her mattress, but now it was my turn to shift around restlessly all night, stuffing spare clothes under my shoulders and bottom to try and get a little comfort, all to no avail. The hammock must have made me soft for this kind of thing, and it was no consolation—indeed, it was rather annoying—to hear everyone else in the shelter breathing deeply, fast asleep.

I never got used to this bed, so during this period of our venture, I took to rising very early—even before Lucho and Marino got up at around 5.30 a.m.—and fanning some life into the glowing embers of the fire, to boil up a kettle. Then I'd wash up last night's supper dishes and perhaps a few clothes, all before breakfast. In the evenings there was plenty of time to wash both dishes and clothes, but we had given up washing anything at night after losing at different times, a T-shirt and two forks, which slipped away unnoticed in the current. A T-shirt and a couple of forks may not sound very important, but they were. Up there it was impossible to get replacements for anything and I learned for myself why the local people—and for that matter, poor people everywhere—took such good care of even their smallest and least valuable possessions.

Lucho was up even earlier than usual the first morning, and disappeared off down the beach, to hunt for turtle eggs. Oliver was moving round a short while later, preparing the mixture for the porridge. He was very fussy about his porridge, and had it all worked out: so many cups of water plus so many of oats, and it came out just the right consistency. I was banned from making the porridge because Oliver said I never counted the cups right.

Lucho came back with a couple of nestfuls of eggs, about fifty of them. This was good news—we could look forward to

one of our favourite lunches. As he turned to face the river he let
out a great shout:

"Look at the view! That should please Señorita Katie."

There, far, far away above the treetops beyond the opposite
bank, maybe one hundred miles away, we later calculated, were
the Andes. Snow-capped, towering through streamers of flimsy
white cloud, the eastern slopes shone brilliantly in the early
sunlight. Katie was up in a flash, hair everywhere, pulling on a
shirt, and we all stood there on the bank gazing at the mountains.
After weeks of oppressive heat and the overwhelming surroundings of the interminable green jungle, those cold, clean peaks
were a strangely uplifting sight. They revived memories of pure,
fresh air, and skiing in crisp powder snow.

However, within half an hour the peaks had disappeared from
view, even though the sky remained clear. Most of the time
there is a low-level haze in the jungle which prevents you from
seeing long distances, even on the clearest of days. We were to see
that wonderful view only once more during our stay at that
camp.

The morning's work was tedious because it took so much
longer to get going with only the five of us. When the system
was actually working we probably got through nearly as much
as with the full crew, but dragging all the gear down to the beach
every day, setting up the tripods, positioning the trough and
the tubes: all this took two hours, and we were tired even before
we started the real work. Oliver had some firm ideas as to exactly
where the jet of water should be played over the stones, and now
that the obstinate Villanueva and the others had left, he could
really start putting them into practice. His theory was that the
water should be played on to the stones just above where he was
working as stone turner, so that the water didn't surge over the
sides, taking some of the gold with it. Katie wasn't used to this,
having learnt a different method when working with Alfonso,
and she kept squirting the jet in the wrong place. Once or twice

we heard loud words exchanged between them above the noise of the water splashing out, the harsh grating of stones on the metal tray, and the chugging of the water pump. By evening we were all really tired and scarcely had the energy left to prepare a meal.

We soon found that there were a lot of animals in our new area, which made it doubly frustrating that we hadn't enough ammunition to do any serious hunting. Each morning we would see fresh tracks around the camp; animals were coming to the water to drink. Often we saw the spoor of a deer, and one evening Katie and Oliver saw the deer itself as they strolled along a dried up section of the river bed, very close to the camp. Several times we saw the curious three-pointed hoofprints of a tapir, and there were too *tigrillo* tracks and jaguar tracks — identified as such by the two boys — which was a little spooky so close to where we were sleeping without walls to protect us. One morning Marino asked us if we'd heard the jaguar calling in the night. We could only say that we didn't think so, because we wouldn't have recognised it even if we *had* heard it. Another time when we were speculating as to the source of a strange hooting, growling sound he told us that it was The Boa — the one that lived in an enormous pile of driftwood and fallen trees at the water's edge not far from our shelters. To this day I'm not sure whether he was joking or very serious. He used to clamber unconcernedly over this heap of flotsam to reach a pool that that was very good for fishing, but we had been told many times that boa constrictors like nothing better than a dark hole among this kind of beached timber. From this time on, whenever we heard a strange noise, we would all immediately chorus: "It's the Barking Boa!" With such simple-minded stratagems we kept ourselves amused and our spirits up when we might be even more tired than usual.

Later I was to hear a story about boas from Wilfredo. He told us that he had once seen a black boa attack a tapir. Now a tapir is a comparatively big animal weighing several hundred pounds, and it is a tough, determined beast. According to Wilfredo's account the boa encircled the tapir with its tail, using its head and upper body to anchor itself to a tree. The snake squeezed and pulled, getting its tail several times around the animal's body, and stretching out like a piece of elastic as the tapir struggled fiercely to break free. Ultimately the snake won, and the tapir lay dead and broken. "There couldn't have been a whole bone left in its body," Wilfredo said. I forgot to ask him whether the snake was actually able to swallow this vast meal.

We had some less disconcerting wildlife near the camp, too. There was a great colony of little, green, screeching parakeets living in the forest behind us. Lucho said if we could catch one of these we could tame it quite quickly, and it would soon learn to talk. But we didn't want to: they were so noisily full of life, flying off in an iridescent green cloud high in the sky, filling the air with piercing screams. Every so often we'd see real, big parrots flying over with their brilliant plumage—reds, blues and yellows—and their long tails trailing out behind them. Somehow these birds looked unparrotlike in flight, and only assumed their "normal" characteristics when clambering over the trunk of a tree and scraping their beaks on the bark.

We settled into a routine of working hard, hunting eggs, and fishing. Once or twice Marino brought back little birds to eat, bagged with the last but one of the shells. The last one was kept for any emergency, "The *tigre*," said Marino. These little birds were speckled brown and looked like wild partridges, though I imagine nothing could be further from the habitat of a real partridge than the jungle. Marino cooked them and yet again we rejoiced at our good fortune in having someone who knew the very best way to prepare anything we could find to eat.

There was very little river traffic. The occasional raft might drift downstream, or a motor canoe, heavily laden with provisions and material for gold panning, with children and chickens

8

crawling over the gear and women preparing meals in the bottom of the boat, with the whole outfit sheltered by a canopy of bamboo or banana leaves, as though the canoe itself was their home.

For our coming Sunday we had plans. We wanted to take the day off and pay a call on Wilfredo, to see if we could buy some fresh fruit, bananas and *yuca*. We didn't know then that our outboard motor wouldn't work, and that we were to be stuck in the vicinity of our camp for the next couple of weeks.

CHAPTER TEN

Sunday visiting

The local supermarket

"How shall we get there then?" asked Katie. It was Sunday. Oliver and Marino were standing over the silent outboard motor, covered in grease, and wearing helpless looks of defeat.

"We've had two days of rain, and the river's high enough. I suppose we could try poling up there," I said doubtfully.

But the look on Oliver's face effectively vetoed that suggestion. Poling was hard work, and Olly knew who would be doing it. He wasn't feeling energetic today.

He turned to Lucho.

"How far would it be on foot then?"

"Probably half an hour," shrugged Lucho. "Maybe a little more."

So we decided to walk. It wasn't far as the crow flies to Wilfredo's place; every so often we'd see smoke from his fires climbing above the trees beyond the next bend of the river. Marino decided to take a lazy day off and stay in the camp, and Oliver himself was in two minds. But after all, as it *was* only half an hour away, he thought he'd come.

"Once we get through this bit we'll be fine," Lucho would say. It got repetitive after about the eleventh time. We kept coming across these difficult patches, where Lucho had to hack through with his machete, and it was slow going. To add to the problem, tall gangling Oliver walking in third place didn't nearly fit into the low tunnel hacked out by little Lucho, so he had to chop out an extra couple of feet at beard altitude, otherwise he would have been almost crawling on his knees.

Lucho took us inland, away from the banks, hoping to find it easier going there. There was no noticeable improvement, and worse still, after we had been walking for an hour and a half we seemed to have lost the sound of the river.

"I'm sure we're heading too much to the north," called out Katie from the rear. What she was talking about I don't know as she gets mixed up when it's just a question of left and right, and I'm no better with greenery encircling me.

My belief in the infallibility of Indian guides was wearing a little threadbare. Lucho himself said nothing, but eventually shinned up a tree in best explorer traditions, to see what he could see. When he returned to the ground he set off again in a new direction, with a visible air of confidence. We all trooped after him, footsore, weary and rather fed up.

We had been going for nearly three hours when at last we heard the sound of dogs barking and cocks crowing. We entered a clearing strewn with felled trees and soon arrived at the top of "Banana Tree Avenue". Wilfredo came out to see what the commotion was.

"Wonderful," he said. "Nice of you to come. You're just in time for a game of football." Obviously not realising that we'd just walked a gruelling three hours from our camp, he hustled Oliver and Lucho over to a clearing behind the huts where a group of men were kicking a ball around. The two boys seemed to shed their fatigue the moment they stepped on the field, and they were soon running and tackling with the best of them. They played on Wilfredo's side, and appeared to be winning, as far as Katie and I could make out.

There was one boy who wasn't playing. He sat on the sidelines looking very miserable, his skin a purply colour, all covered in scabs, his wrists and ankles swollen. He was obviously in great pain, and we wondered vaguely what was wrong with him.

The game went on for ages, punctuated by frequent triumphant shouts of "Gol!" but with no attempt to keep score or impose rules that we could see. At last the players stood panting and dripping with sweat, and agreed unanimously that they'd all had enough. They went trotting off to the river for a swim. Wilfredo came over and told us to stay for lunch. We felt a bit awkward about this but Wilfredo waved aside our protests.

"We'll have a chicken today then," he said. He vanished into a hut and reappeared a moment later with a handful of rice, which he proceeded to scatter on the ground around us. All the chickens in the yard converged on the rice, flapping and clucking.

"Which one of these hasn't got any chicks to look after?" he shouted.

"That old white one over there hasn't got any now," said one man. "The vultures got the last one yesterday."

At a word from Wilfredo three men surrounded the unlucky bird, and its earthly career ended with a brief squawk, as one of them wrung its neck. This man set about plucking and gutting the carcass, throwing the offal to the dogs who made short work of it. Soon a chicken stew was bubbling on the open fire, while one of the men cleaned some rice. He pulled handful after handful out of a big bag, letting the loose grains stream through his fingers and blowing vigorously to remove bits of husk. It took him ages to clean enough; Wilfredo had six men, and there were four of us, so we were eleven for lunch that day.

Someone had killed a snake that morning, and it was brought from the field for us to inspect. The thing was about three feet long, brownish-beige with dull vaguely zig-zag markings and was quite definitely poisonous, judging by the respect with which the men treated it, even though it was very dead. They carried it at the end of a long stick, and when they turned it round to look at it they did so by poking it with another twig. There was no question of touching it with their bare hands.

"What is it?" we wanted to know. So far we'd been fairly lucky, having seen only the snake that our boys had brought back at One Bend Camp. That one hadn't been dangerous.

"It's a *jérgon*," they said.

Later we discovered that *jérgon* is the local name for the fer-de-lance, one of the three most poisonous snakes in the Amazon (the others are the bushmaster and the coral snake). Normally, we were told, the snake—a member of the viper family—would wriggle off in a hurry if it heard something coming. Only if one were to catch it unawares and virtually tread on it would it strike, they said. Nevertheless, the sight of this lethal creature so close to our own camp gave a fresh lease of life to our slightly neglected rule of always wearing boots, with thick socks and trousers tucked in. A single layer of clothing would reduce the penetration of a snakebite, and might prevent it from breaking the skin at all. The real danger came if the snake managed to get you in an area of exposed skin, and inject all its poison. Wilfredo rightly said that snakebite deaths were fairly rare—much rarer than actual snakebites though they did happen occasionally.

Wilfredo cautioned us to be very careful when collecting driftwood for the fire; always give the branches a kick first, he said—there might easily be a snake dozing underneath. The *shushupe*, or bushmaster, he told us, was the one to really watch out for. This member of the viper family is one of the largest venomous snakes in the world, six feet and more in length. Unlike most snakes it's very aggressive, perhaps because of its size, and will even stalk a person, waiting for the right moment to attack—that's what Wilfredo said, anyway.

We often teased Wilfredo about his age, saying he must be having us on—he couldn't really be forty-one.

"You know," he said once. "I eat all the parts of a meal that you people discard. I eat a fish's head, I chew bones and bite them to suck the goodness from the inside of them. That's where all the goodness is. If we kill a boa I eat that too. Any animal that sheds its skin is good for you to eat. It gives you a new lease of life. That's what keeps me young."

We had some coffee and then decided we should be getting back. Luckily for us Wilfredo insisted that we should accept a lift home in one of his canoes. That wasn't the only thing he insisted on: he loaded us up with bananas, *papayas*, *yuca*, *ají* and maize, and he flatly refused to take any payment. We felt very uncomfortable about this; we'd arrived unannounced for lunch — not that there had been any way of letting him know we were coming — and eaten one of his precious chickens; now here we were, walking off with armfuls of his produce, the fruits of his labour. But Wilfredo's expansive generosity and hospitality were impossible to decline graciously, so we thanked him gratefully and accepted.

Just as we were about to leave we were given the opportunity to do one small favour in return. And at the same time we learned the answer to a question that we had been pondering earlier: what was wrong with that boy we had seen looking sick and miserable at the side of the football patch? The boy was apparently a new-comer to the jungle, and had been almost eaten alive by the mosquitoes. He had taken to scratching the bites, thereby causing open wounds which had now festered and turned into a general skin infection.

"It's because he never washes," said Wilfredo bluntly.

Anyway, Wilfredo had sent for some penicillin to Puerto Maldonado, and it had arrived. But he had nothing to administer it with. Did we have a syringe? We did, and of course we would be happy to give it to him. Lucho volunteered that Marino had worked as an orderly in a hospital in Cuzco, so we even had someone to give a proper injection. We were glad to help this doleful and pathetic-looking boy. He'd left his home in the high

Andean town of Puno and come all this way, only to find within a short time that he was too sick to work. Because of this he was earning no money, and in fact probably owed money to Wilfredo for his food. Not a happy situation.

Wilfredo said he would send the boy down next day for his injection, and mentioned that he himself would be going to the Banco Minero's camp in a couple of days' time to stock up on provisions. He agreed to pick me up on the way and give me a lift there and back, as we had some shopping of our own to do.

In the canoe we covered the journey back to the camp that had taken us three hours on foot in just ten minutes. We found Marino peacefully fishing near the camp. Katie was a little worried about him. He'd been very gay at One Bend Camp and he was still cheerful and very helpful. But he preferred going off on his own to joining in whatever we were doing. When we asked him if there was anything wrong he would smile politely, look into the distance and say that his heart was sick.

We were very excited with our new food supplies after the dreary things we'd been eating for the last few days. There were about a dozen *papayas*, and until they ran out we had this succulent melon-sized orange-coloured-inside fruit, with every meal. Katie fried the bananas, and we tried them ember-baked too, along with an odd potato-like vegetable called *uncucha* which we hadn't tried before. Our new abundance of food caused an unexpected problem, a plague of mice. They ate through anything we had to wrap the food in — plastic bags, sacks and cardboard boxes — so there was nothing for it but to let them get on with it and hope they'd leave some for us. At our last camp the men had slept with the food stacked on boards around their heads, claiming that this kept the mice at bay. But as we'd shifted the food prior to leaving I'd seen mice scattering in several directions.

The following day as we were working hard on the gold operation four of Wilfredo's men arrived in a canoe, accompanied

by the bloated sick boy, whose nickname was Gordo — Fatty. I
went with Marino to hunt out the syringe. When we returned
Wilfredo's men were hard at it, filling up wheelbarrows and
running them up the plank. Oliver was working like a zombie at
double speed to keep up with the unprecedented inflow. I didn't
want to interrupt this fantastic show of hard work, so I took
Lucho aside and told him to send them up to the camp when
they'd had enough.

I had made a huge rice pudding, that morning, and had a big
cauldronful of it steaming over the flames by the time the men
knocked off. I sloshed two very full ladles on to each plate so that
they were brimming over, and watched the men scoff it all down
faster than I would have believed possible. They all accepted
second helpings of equal magnitude, and afterwards got up to
leave with no sign of having overeaten. Probably there was
another meal waiting for them when they got back.

We all looked dubiously at Gordo as he shambled off after the
others, and I dropped his plate and spoon into a pan of boiling
water. Whatever was wrong with him, it wasn't pretty, and for
all we knew it was highly contagious. He came to our camp again
several times, and even let us know via Marino that he wasn't
happy at Wilfredo's and wanted to stay with us temporarily
until he could get a ride to Puerto Maldonado. "No way," said
Olly using one of his favourite expressions, and the subject was
dropped. Gordo did improve a bit but what he really needed was
to get back to the Altiplano.

There were five cartridges left, so we decided we could lash
out and have a day's hunting. We wanted to see what the jungle
was like around our new camp, and it was about time we had
some meat. We struck out into the forest behind our clearing,
with Lucho leading the way, and Marino bringing up the rear.
Soon Marino made some excuse to leave us — as he always did
when he had the chance — and faded into the jungle. As the rest of
us continued we saw a group of monkeys. They chattered and

screeched, seemingly indifferent to our presence, though they kept their distance. I could just catch a glimpse of them from time to time as they swung from one branch to the next. But I found these and any other animals we encountered in the jungle frustratingly difficult to see, even when they were pointed out to me. Katie was much quicker than I to spot them, and so was Oliver.

Later we heard a shot quite close by, and we called out to let Marino know where we were. Moments afterwards he appeared through the undergrowth, holding aloft a small partridge-like bird. We saw no more game that day.

Wilfredo had promised to collect me early next day on his way down to Camp Colorado. It was lucky I had got into the swing of rising very early by now, for he appeared at the river bank in his dugout canoe just as I finished dressing. I grabbed my shopping list — though experience had taught us never to be optimistic about finding anything we wanted — and trotted off to the water's edge.

Mist was still lying low on the fast-running river, the dawn smelt fresh, and the birds and animals seemed to have only just woken themselves up. Wilfredo sat perched at the very end of the canoe, I squatted in the middle, and Marco, one of Wilfredo's workers, knelt in the prow.

There were only two paddles, so I relaxed comfortably with a camera in my hands, hoping that we might spot some unusual animal as we drifted downstream. But, of course, that would only have happened if my camera had been tucked away safely out of reach. Two ducks flew over, but by the time Wilfredo had his gun up they were out of range.

It took us two hours to get down though we couldn't spend too long once we were there, as it would take us three times as long to get back with the canoe heavily loaded.

Feliciano who was in charge of Camp Colorado, spotted us coming in, though, and came to the bank to greet us.

"Where's your canoe?" he asked me.

"The motor's bust," I said. "I had to come down because we're really low on supplies. We've only rice and beans left."

Feliciano grimaced. "We haven't got much else ourselves," he said. "Over these holiday periods everyone seems to forget that we have to go on living up here. I suppose they can't find anyone to bring supplies up, the *motoristas* are probably all too drunk."

We followed him into the store. It seemed quite well stocked with supplies, but they were in the same situation as us—they had all the boring basics and nothing interesting. Wilfredo brought out a small piece of folded paper and placed it on the table.

"Thirty-six grams," he said. From another pocket he took out a shopping list. "I want provisions to the value of that gold— let's hope they last three months."

Feliciano unwrapped the tiny packet and delicately placed the yellow lump on the scales. We all peered as the needle swung back and forth, finally settling at 35·89 grams.

"Okay, Wilfredo? We're buying at 135 soles per gram—you get about 4,800 soles."

Feliciano was trusted by the miners. His job was to encourage them to come in and sell their gold to the Banco Minero, rather than to hoard it or try and smuggle it across the border. Not only was the Bank paying fair prices for gold, it was also selling provisions and equipment at remarkably low prices, undercutting the traders who had formerly held sway on the river.

The two men worked out their sums, and Marco began hauling out sacks to load aboard the canoe. Meantime I browsed round among the bags and cardboard boxes, looking for anything that might interest us. I found some tins of milk and some packets of pepper and comino, all basic things that we needed badly. In the way of luxuries, the only things I uncovered were a few tins of sweetened condensed milk, something Katie had grown very fond of.

"Feliciano, we're right out of sugar, cooking oil and coffee. Have you got any?"

"Not much, but if you really need some I'd better give you a little from our private supply. We're very low ourselves. They did say on the radio last night that we should expect a canoe-load of provisions in four days' time. But who knows what that will contain."

Feliciano weighed us out a few pounds of precious sugar, measured out some cooking oil, and handed me a small tin of coffee. The meagre results of my shopping trip didn't take up much room in the canoe compared with Wilfredo's purchases. These completely filled the canoe, leaving little room for the three of us and leaving the canoe so low in the water that Lloyd's would never have insured it. Marco had laid strips of bamboo six inches above and lying across the bottom of the canoe, for the sacks to rest on so that they wouldn't get wet if we shipped a little water. Everything was stacked in perfectly, like pieces of a jigsaw puzzle and the entire load was covered in plastic sheeting.

"That's in case it rains," Wilfredo said. "I'd hate to get this lot wet. It goes bad so easily even with the humidity. We've had it if water gets in."

Feliciano then told me that he thought the only trouble was that the drum of mixture for our canoe motor was too rich in oil, but as he had no extra petrol to give us we would have to wait until our boat driver came up.

We set off just after 10.00 a.m. It seemed like midday already, with all that we'd accomplished, and we felt like taking a break. But we had a long and tiring journey ahead of us getting back upstream. As we slid away from the bank Feliciano called that he would send us news if he received more provisions, and then Wilfredo and Marco started the hard work of poling upstream. They each had bamboo poles at least ten feet long which looked so unwieldy I wondered how they didn't keep falling over them. The river was very wide but, I realised, surprisingly shallow. Only in a couple of places were they unable to touch bottom and then they would both dump their poles and paddle furiously, before we lost too much headway.

After watching their performance for over half an hour I finally plucked up enough courage to pick up a pole and start copying Marco in front of me. We soon stopped for a tin of tuna fish and dry miners' biscuits.

"If you're going to pole on the way back," said Wilfredo, "you'd better eat as much as you can and get up plenty of strength."

We carried on poling all afternoon dipping our hands into the river every so often as we pushed; Wilfredo said this helped prevent blisters – but the advice was a bit late for me; I had them already.

We finally pushed into camp around 4.30 p.m. exhausted and hot. I went straight to the river to have a good wash while Katie and Marino prepared a meal for everyone. I overheard Wilfredo telling Lucho, "She's been working so hard today, she's going to sleep well tonight" – nodding his head in my direction. That made me smile quietly to myself. He was right – I was completely whacked, and couldn't wait to flake out under my mosquito net. But there had been an ulterior motive behind all my efforts that day. I wanted Wilfredo to take me on one of his hunting outings, but I felt I had to show that I could join in if there was work to be done. Previously he might have protested that such a trip was beyond my endurance – but after today's outing I was to get my own way.

CHAPTER ELEVEN

Indian visitors—a dozen of them

A hunting trip

THE NEXT FEW DAYS were rather miserable ones. The weather was overcast and cold and though we made an attempt to work we were drizzled out and gave up. As usual, bad weather meant poor hunting and fishing. There was nothing much to do.

So we were quite pleased one morning when we received no less than four canoe loads of visitors. The first three canoes arrived from upstream. They were Mashco-Amarakaire Indians from a village located a day and a half's canoeing upstream. There were twelve of them, all men, and though they were dressed in the now standard dress of shirts and trousers they had very striking features and, as Oliver pointed out, enormous and very wide feet. Only one of them spoke Spanish, and he not very well. Among themselves they conversed in the language of their tribe. Neither Lucho nor Marino understood a word of it.

Their spokesman explained that they were bound for Camp Colorado to get in supplies of beer. They had a big fiesta coming up, he said, and intended to buy many crates. But the fourth visiting canoe, which arrived at that moment, was to dash their hopes. It contained a little man who had gone down the previous day on a similar mission. He reported that there was no beer to be had, either at Orosco's shop or at the camp.

The Indians accepted this news without visible signs of annoyance or disappointment, even though this meant that they'd be having rather a dull party, and they'd come a long way for nothing. But they all looked rather cold and hungry so we offered them a meal which they accepted. It was customary we'd learnt to invite any visitors to a meal at any time of the day. Marino measured out rice for eighteen people, thirteen visitors and five of us, and put it on to boil. This was the biggest meal we'd ever cooked for anybody, but luckily Marino had a good eye, and knew exactly how much rice eighteen people would eat. When the rice was ready he mixed in several tins of tuna fish, and the meal commenced.

There were enough plates to go round for the first sitting of our visitors but eating irons for only nine of them. We quickly realised that this didn't matter at all, because the Indians were used to eating with their fingers. When they had finished they all trooped off in unison to the river to wash their plates—good guests! After the meal their spokesman said that they had animal skins to sell—would we like to buy any? We weren't really interested, and said—truthfully—that we hadn't any money. No matter, said the man, we could trade for gold. Then we had to admit that we didn't really want the skins. Now if they had bead bracelets and necklaces ... No, he said, they hadn't any beads. Then there was a general shaking of hands, and our visitors departed on their long haul back up the river.

That night at nearly midnight I awoke with someone shaking my shoulder. It was Katie. She was in tears, and doubled up with agony.

"Anna, I feel so awful, I've got such a terrible pain in my stomach."

"Which side is it on?" I was thinking of appendicitis, but I couldn't remember where the appendix was, so I had to wake Oliver.

He was up in a flash. "It's on the right side," he said. Katie said the pain was everywhere, but particularly in the middle. We decided it couldn't be appendicitis.

"Did you eat anything that we haven't eaten?" we asked. It seemed that she hadn't.

We got out the medicine bag and hunted around for some Milk of Magnesia. Everyone was awake by now, and Lucho suggested she should try some tea made from comino. "It's very good for upset stomachs," he assured us. While Lucho revived the fire and put the kettle on I took Katie for a little walk along the beach. She'd been sitting all cramped up, and I thought that the movement would do her good. But all the stuffing had somehow been knocked out of her, and even leaning on my shoulder she couldn't go very far. After drinking the comino tea she began to feel a little better. By morning she wasn't in too much pain, though she was still as pale as a sheet and very weak. She stayed in bed all day while we put in some gold work on the beach.

Next day she was up, but still very weak, and wasn't nearly strong enough to work again. But when we came back for lunch we found her in tears once more.

"I was afraid those hot peppers Wilfredo left us would go bad, so I cut them up to cook and bottle them, and now my hands are on fire." She *was* miserable.

"But, señorita, you must never cut those peppers up. They are very, very hot," said Marino, rather unnecessarily at this juncture. We wished that for once he would shut the stable door before the horse bolted; for several days there'd been talk of cooking those peppers.

Nothing Katie did would rid her of the terrible burning in her hands. She tried every cream, rubbed cooking oil all over them, blew on them all the time, and ultimately squatted at the river's edge holding her hands in the cooling stream. It wasn't until nightfall that the pain began to ease.

Next day turned out better. Katie was back to normal, it was a sunny day, and it was a Saturday, the day we usually clarified

Oliver and Lucho panning
Gold workers in Rio Caichive region

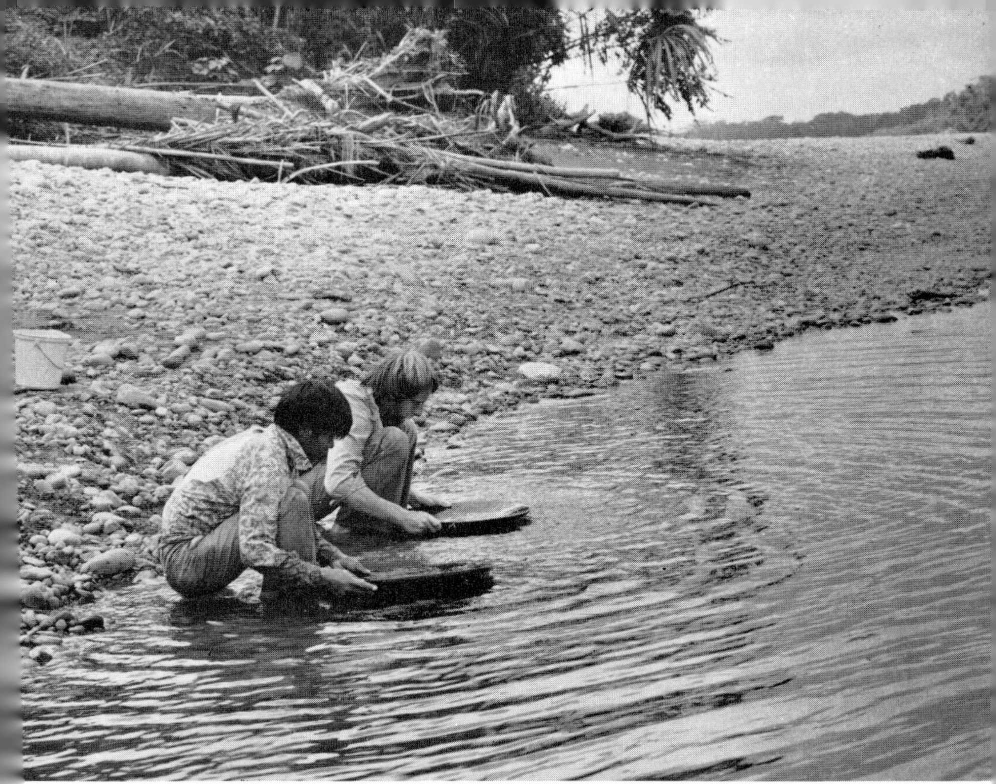
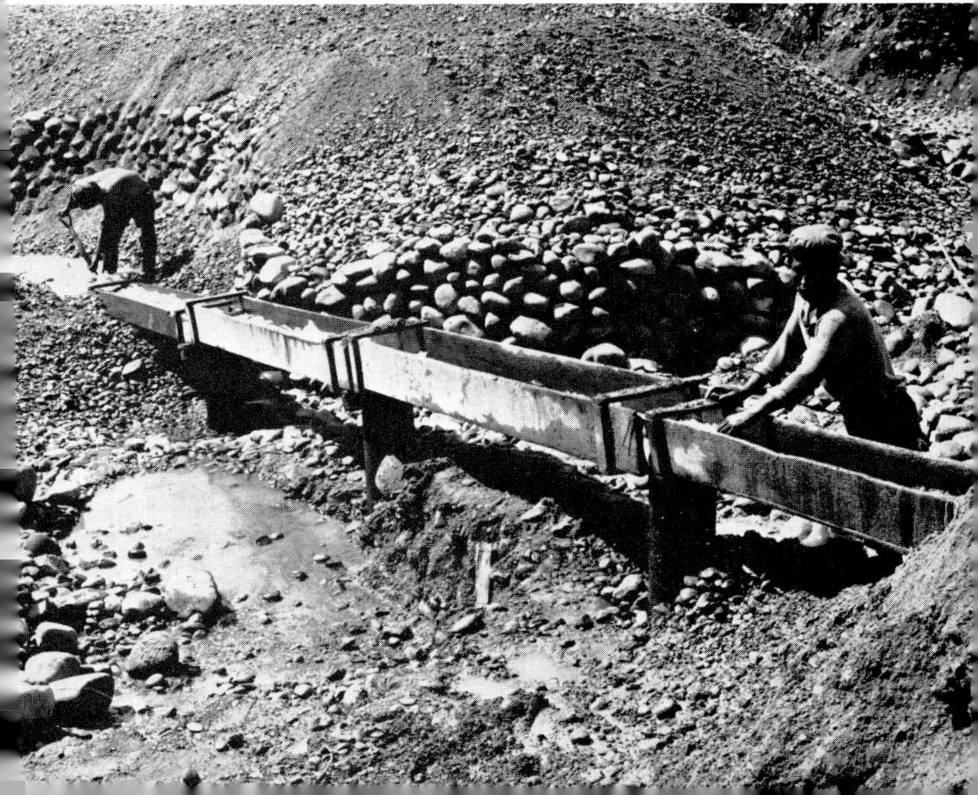

our accumulation of black sand, to reduce it to gold. This was a job just for Oliver and Lucho, so Marino went off hunting, and Katie and I set out to do some fishing.

When—like today—we didn't have any suitable bait, catching our meals was a muddled three-stage process that was fun, but time-consuming. First we had to catch some of the tiny two-inch tiddlers that swarmed in the shallows. With a fine shrimp net we could have caught all we needed in a few minutes. As it was it took ages, using the dirty cooking pots as a baited lair, and bringing them out in ones and twos. We would put these minnows on our small hooks to catch slightly larger fish, about six inches long. Only when we had these could we go after the big stuff.

It was quite late in the afternoon by the time we had enough bait and the other two had already finished clarifying the gold, which they'd put under my pillow—it was in a little pill bottle I had, and weighed about twenty grams.

While Oliver stayed behind to read, Lucho, Katie and I set off paddling the canoe downstream to a deep pool we knew of. Katie caught the first fish, vicious, black and sharp toothed and about eighteen inches long. Lucho said they were delicious to eat though he couldn't remember the name. Later I had an enormous tug on my line which nearly pulled me out of the boat. Lucho hauled in the fish for me, a *sabalo* he called it, more than two feet long. Now we had plenty for supper, and it was nearly dark so we got ready to paddle back upstream. Just as we were leaving a voice called out to us from downstream, and moments later a canoe pulled alongside ours.

"*Señorita*, I have a letter for you from down at the Bank," announced its occupant, a gold-toothed prospector living further upstream. "It's following in another canoe behind me. Do you think the four of us could stay the night at your camp?"

Lucho leaned over and whispered that this was standard jungle courtesy for anyone who couldn't make it to his own camp before dark.

A difficult passage, Rio Colorado
The daily toil upstream

9

"Well, of course," we said.

Katie and I were excited at the thought of letters; Katie especially—it was her birthday next day, and she would have loved to receive some news. I was hoping it would be something for her, because Victor had not yet returned from Puerto Maldonado, and I had no present for her. Soon the second canoe arrived, and the letter was duly delivered—to me. But it was an anticlimax all round; just a courtesy note from Feliciano to say that he hoped we were all right, and that our new *motorista* would be arriving in two or three days. This last piece of news we'd already heard on the previous night's radio programme.

Our visitors presented us with a duck they'd shot and we all shared a supper of our delicious freshly caught fish. When we turned in the men borrowed one of our lamps, and slept curled up round the fire with the lamp burning and a radio playing all night. As they didn't have mosquito nets, they explained, these tactics were necessary to frighten off vampire bats. That was the first time we'd ever thought of or heard about vampire bats. But then we'd never slept without mosquito nets. It was also the first time we'd heard of radios being used as an anti-bat measure.

Our visitors left silently in the early hours of the morning, for they were gone by the time we woke at five o'clock. I thought it was very strange, because we had prepared the duck especially for them to have a good breakfast before they left. But Lucho said these men were used to travelling early, and preferred it.

No matter about the duck anyway. It was Katie's birthday and we were delighted to have it all to ourselves. We had a tortoise, too, that Marino had found on his hunting expedition the day before, and there was plenty of fish left. Today we had arranged to visit Wilfredo and celebrate Katie's birthday there with a good feast and a special football match to mark the occasion. Before leaving I made a big bowl of dessert from our final packet of Heinz Erin Creme Caramel.

We loaded everything into our large and unwieldy canoe and poled off upstream, with me clutching the bowl of dessert to keep it from slopping everywhere. We arrived after midday, but there was no sign of any lunch cooking yet. The first scheduled event, explained Wilfredo, was the football match. Lunch could wait until they'd finished playing, whenever that would be. It was eventually to be four o'clock.

Marino had come with us this time, and turned out to be a first-rate football player. Our three boys played with Wilfredo against the rest of Wilfredo's men, and the match went well for a while. Then Oliver came limping off the field, looking miserable and cursing his luck.

"What's the matter, Olly?" called out Katie.

"I've pulled all the muscles in my thigh," he said dejectedly. "That's messed up the whole game now. The numbers are wrong and they've been looking forward to this match all week." He hobbled miserably away.

As our football match went on, the radio was blaring away with the commentary of another football match — one of rather more national importance. Peru was playing off against Chile for the World Cup qualification, after drawing on aggregate from two previous matches. Every so often someone would fall out of our game to come and see what the score was. As it gradually became clear that Peru was losing, the men stopped coming to listen and played on, taking more interest in their own game.

When we finally got to Katie's birthday meal, the wild duck turned out to be a rare treat, absolutely delicious, even though we had no orange sauce. Wilfredo was terribly upset, though, that there was no beer or wine.

"How can we celebrate Señorita Katie's birthday without a good drink," he said.

"Well, I don't think it's that important to Katie," Oliver said. But Wilfredo couldn't be persuaded that it didn't matter and looked upset all day.

Oliver's legs were no better and he could only walk very stiffly. Wilfredo had some Charcot, a special liquid rub, which was good for muscles and Oliver disappeared into Wilfredo's room to apply "the cure". He came out full of admiration for Wilfredo's room.

"Hey, Wilfredo, it's amazing in there," he said. He turned to Katie and me. "He's even got a proper bed with a mattress." That was something we'd never seen in these parts. "It's all set up like a shop, the provisions are laid out along shelves — tins and tins of food. He's much better stocked than old Orosco. There are some animal skins, too."

Wilfredo then fetched out some of the deer and ocelot skins to show us. He told us he had several ocelot traps in the area and when he went to Cuzco he would sell the skins. It was a sideline to supplement his income from gold. Wilfredo was very proud of his skins and wanted us to take colour photographs so we could send them to him. It dawned on us afterwards that it would be difficult to send them to him. Wilfredo had no address and there are no postmen within a hundred miles of his place.

Before leaving we made arrangements to film Wilfredo and his men working gold on their beach. He had a different system from ours because he was working without a water pump. We were expecting our new *motorista* to arrive the following day, so we could start moving a little more freely, we hoped, during the coming week. Even though we had no chocolate birthday cake for Katie we'd eaten pretty well, and back at the camp we made do with a hot cup of tea before going to bed.

Lucho and I were up early next day because we'd decided to go and collect our new boat driver at Camp Colorado and I wanted enough time to collect some turtle eggs first, as a present for Feliciano. They had been left with neither gun nor canoe for the past two weeks; we had the gun and Victor had the canoe. So they hadn't had much chance to hunt for fresh food. We had a special beach where we knew there were always turtle eggs providing the weather was warm enough. Today we came back laden.

I had quite a fright while we were hunting for the eggs. I was happily digging away, when Lucho caught my eye, grinned and pointed somewhere a little way ahead of me. There, not twenty feet from me, was a four-foot alligator, staring me right in the eyes. It quickly turned round and scuttled away, diving into a pool before I had time to blink.

"Those are the little beasts that keep eating the bait off our fishing lines and making us think a big fish is nibbling," said Lucho.

Katie had been disappointed when she'd seen her first alligator. She'd expected it to be bright green, like a dragon or something. In fact it was a dull greeny-brown, much the same colour as its surroundings, like many other animals we saw. This was one of the reasons why I was never quick enough off the mark with the camera where animals were concerned. As an excuse I have a suspicion that ninety per cent of the wild animal photographs one sees in magazines are taken in special zoos and not way out in the wilds.

We had enough eggs for Feliciano and ourselves, so we all set off to paddle downstream to Camp Colorado. Though we got away with it, it was, in retrospect, a foolhardy thing to do. Our canoe wasn't nearly as low and manoeuvrable as Wilfredo's. It was much heavier with high sides and we only had one paddle between us. Marino stood at the front using a pole to help with the steering.

We made it, rocking and bumping, through two sets of rapids, barely avoiding a collison with a gigantic tree. Then we spotted an orange-painted canoe identical to ours, making its way upstream. So they were even bringing our *motorista* up to our camp for us.

Somehow we managed to push and coax our canoe into the bank, and they drew up alongside. We had been wondering who they would send us, hoping anxiously that it would be someone who would fit in with our camp. It was a pleasant surprise to see that our new driver was Juan, one of the two men who had brought us upriver in speedboats when we first left Puerto Maldonado. Juan was all right, a good-humoured, unflappable kind of person, slowish, but doggedly persistent in finishing anything he had started. Getting him was a piece of luck.

There was no real reason to carry on downstream now, but Katie had set her heart on a day out on the river. Juan got our motor going after siphoning some petrol from the other boat's tank and tinkering for about five minutes. We had no further trouble with it. Oliver sat there glumly, muttering about Katie always getting her own way, and we carried on downriver to visit Feliciano and collect a few more provisions — and have a day out on the river.

The boys now had to sleep four across the width of our shelter, and were very cramped. Meanwhile Katie and I continued to lord it at the other end, reclining though still on our uncomfortably hard bamboo beds. Olly and Marino put in a little time setting up extra awnings to make sure there was still enough cover for all of them.

In the morning we had planned an early start for filming Wilfredo's men working his beach, but Katie just couldn't make it early, much to my fury. I'd wanted to get the filming over before the sun was high but it was only after we'd finally set out and I'd had a chance for my temper to cool off that I was able to laugh with everyone. It *was* rather absurd to start fretting about the clock out here in the wilderness. But then I was a little bit cross all the time during this period because we'd had no news of when my brother was arriving and we wanted to go upstream, so we'd been going through fruitless agony trying to tune into the Bank's programme only to find that either there was no message or we couldn't get the correct wavelength.

As soon as we arrived we saw that Wilfredo had his men really well trained. They were practically running up the plank with their wheelbarrows and they never seemed to take a break. Some men dug as others loosened the earth with picks, and Wilfredo himself worked away feverishly, forever taking samples, determining the working depth and deciding which sections of the beach to excavate. He did his sampling in a matter of seconds just swishing round some material in a shovel, whereas we'd made rather heavy work of it using a panful every time. It's practice that counts, he told us. As Wilfredo had no water pump his trough stood in the river and one man had to work away tipping buckets of water over the stones. The diggers had a lot further to push their barrowfuls than we did because of this arrangement, but they were well organised. As we were leaving one of the men was making a fire on the beach to heat up their lunch, so that they wouldn't waste time trooping back to camp.

Wilfredo said he was expecting to get about three grams per man per day from that beach. Now we felt we had seen a real gold operation. The efficiency and co-ordination of Wilfredo's set up made us feel very sheepish about our own amateur efforts.

We decided to spend the whole day filming, as we were finally more mobile with Juan to drive us. He took us downstream to visit a group of men working for old Orosco the storekeeper. Orosco's men were all *serranos* — men from the sierra. They were clinging to their traditions even to the extent of wearing their knitted wool caps with earflaps for chilly Andean nights, instead of the more practical wide-brimmed straw hats. They were working like *serranos*, too, never stopping, barely looking up, until we asked them to line up so we could take a picture of them.

On the way back to the camp Oliver and Lucho stopped to do some tryouts on a beach though without much success — hardly any gold, they said. Meanwhile the rest of us began to fish; we never went anywhere without our lines and the gun. We caught five fish, each weighing about two pounds. I hadn't seen this kind, and when we cooked them later they turned out to be delicious, with few bones. Actually the only meal I ever really couldn't stomach was stewed parrot, which we had later on.

What we didn't manage that evening we fried up for next day's breakfast. Katie was not doing too well in the food line, though, because she disliked fish and would never eat it. Since it was one of our main staples she was really missing out. All she had to cheer her up was a little condensed milk.

That morning I had arranged to go hunting with Wilfredo. I set off early with Juan, who was going to drop me off at Wilfredo's camp. It was really a fine feeling to be mobile once again after being stuck in one place so long with our recalcitrant outboard motor.

As usual Wilfredo had heard our motor, and knew we were coming long before we appeared. He was there waiting on the bank as we arrived. Juan left for the camp, and Wilfredo and I transferred to a smaller canoe of his in order to cross to the far bank at a spot higher up the stream. There were two stretches of fast-running shallows to cross on the way. The river was very low, barely covering the rocky bed in places, and I feared we would run aground and be turned over by the current.

Wilfredo noticed my nervousness. "What's the matter, do you think it's too early for a swim?"

"I don't mind swimming," I said. "It's my camera that doesn't like it."

During all the time I knew him, I can never remember Wilfredo having been at a loss for a pleasant phrase, an amusing or witty remark. Never were they malicious either.

We pulled into the bank and tied up to a log. "I've made quite a few paths into the forest from this place," he said, "but today we'll get off the beaten track. This is a good area for that—it's not too thick, and there are plenty of animals."

I was a little nervous, and very eager to please. Wilfredo had the reputation in this district of being a first-rate hunter, and I felt it was a privilege to be with him just then. I didn't want to get in his way or make too much noise, so I set off following him, trying to put my feet exactly where he had put his. This proved to be unworkable as concentrating on the ground made me keep running head on into low, thorny branches, and in any case Wilfredo was loping along at such a pace that I could barely keep up with him even moving normally.

It was a beautiful day. The heat we felt on the open beaches didn't penetrate here. In the forest gloom it was almost cool, and the mosquitoes were nowhere to be seen. Often we heard birds calling and monkeys chattering without actually seeing them, and then, quite suddenly, we walked directly beneath a whole family of large reddish-black monkeys. They fled, swinging and leaping through the treetops, howling as they went.

"Those are called *makisapas*," said Wilfredo. At that moment he dropped on one knee to examine the ground. I could see some tracks there in the soft earth.

"*Chancho* — wild pig," he said. "Lots of them. And they're very fresh tracks."

"Now these here," he pointed over to one side where there were more marks, "are deer tracks. But those are a day old."

Being with Wilfredo was an education. Every time he saw something that might interest me he would beckon me over and explain it all in a low, quiet voice. He pointed out wild cotton trees ("but that one's past its prime for cutting") and cacao, the plant for cocoa. He showed me enormous cedars and mahoganys. "You see the way it splays out into blades at the base of the trunk? That part is ideal for making wooden gold pans."

Once we stopped for a rest. "Aren't you tired?" he asked.

"No," I replied. And it was the truth, even though we'd come a long way and done it fast. This was much more fun than working all day on a burning hot beach.

"Well, I don't know about you," he said, after a pause. "Me, I'm always shattered after a day's hunting like this. I have to concentrate so hard, my eyes have to be everywhere, all my senses have to be alert. I can never rest."

Something I had noticed about Wilfredo this morning was that he had taken to chain smoking, even though I had never seen him smoking before. I asked him why.

"It's for the snakes," he said. "Snakes hate cigarette smoke, so I get through a couple of packets when I go out hunting. Do you want one?"

I've never smoked before or since that day, but at that moment I accepted Wilfredo's offer with no reservations at all. It seemed that smoking wasn't *always* bad for your health.

We pushed on. Many times the only way to pass difficult patches was to scramble along the top of a fallen tree. Sometimes one of these would be acting as a bridge over a small creek. The tree trunks were knobbly and slippery, and I had vivid recollections of doing balancing acts on the bar in the gym at school. Only then there'd been a soft mattress to fall on to instead of a dirty, muddy, creature-inhabited stream.

Once there was a peculiar continuous piercing noise.

"*Trompetero*," whispered Wilfredo. I thought perhaps he was going to try and shoot it, though I had no idea what a *trompetero* was. I stopped so as not to interfere with his stalking, but he turned and beckoned impatiently for me to follow. He picked his way up a steep bank, and I scrambled after him, managing to slip and fall, not once but twice. That's sure to have scared it off, I thought. But Wilfredo began to imitate its call, and the *trompetero* responded. The "conversation" lasted several minutes, with Wilfredo creeping silently forward all the while and myself clumsily bringing up the rear. In the end there was a sound of something crashing in the undergrowth, moving away from us at great speed, and then silence. Wilfredo let out a sigh, then he relaxed and shrugged.

"I'm sorry, Wilfredo. I made too much noise," I said.

"No, no. I lifted my head and it saw me, it wasn't your fault."

"Wilfredo, what is a *trompetero?*" I asked. It turned out that it was a large black bird, good to eat but difficult to hunt.

"I don't really bother with small game," Wilfredo said, as a flock of birds flew over. "There's no point in killing small animals when I have all those men with huge appetites. I haven't the time or the cartridges to spend on anything less than big game."

As we continued the only noise to be heard was the occasional cracking of twigs as we set our feet down. I was getting a little better at moving through the forest. All of a sudden Wilfredo was moving ahead much faster, yet even more quietly, his head darting from side to side and then snatching a quick look at the ground to check those same pigs' tracks that we had been following for the last hour or so. Evidently he had heard or sensed something. Suddenly he stopped. "*Venado* — deer," he whispered. He aimed with his gun. "Do you see it?" I shook my head, and he fired. Then we listened. There was a small shuffle — a light thud, then there was no sound.

"It's fallen," he said at length. We walked over to a patch of bushes about thirty yards away and found it lying there dying. I remember feeling a burning behind my eyes; it was beige and so pretty.

"I couldn't get it in the head," he said casually. "I only saw just the white tail flicking there between the leaves, so I had to go for the body. Pity. It spoils the meat a bit." Then he added, with surprising gentleness, "Poor little thing. Why didn't you manage to escape?"

Wilfredo told me to wait and said he was going off to look for something to use to truss up the animal. He disappeared for over half an hour, though all the while I could hear him crashing about in the undergrowth, whistling and chopping with his machete.

As he came back he shook his head. "I just can't find any roots or bark that are strong enough. These will have to do until we find something better."

He was carrying lengths of hanging creeper about half an inch thick—the stuff Tarzan swings on. With these he tied the deer's feet together, leaving a long loop hanging free for him to carry it by. Then he hauled the dead animal on to his back, grunting a few times as he settled its sixty-odd pounds comfortably on his shoulders, and set off, calling for me to follow. He set a rapid pace, seemingly faster than ever.

"We've a long way to go to get back," he announced as he strode along. "We've been going for four hours, and it'll take us at least that long to return."

We stopped several times. At one resting place he sat on a log and pointed near where he was sitting. There, crudely carved in the wood, was a blackened but legible W.

"I must have carved that at least four years ago. I haven't been down here this far since then," he said.

All the jungle looked the same to me. I would never have found my way out or been able to follow our pathway, so it amazed me that he could remember a particular tree trunk from four years ago. He asked me if I could point to where our camp was, and to his place. I really hadn't a clue, and it was pure luck that my guesses were no more than forty-five degrees out. At one point a bit further on Wilfredo motioned me to stop and listen, and there, very faint in the distance, was the sound of the little petrol-engined water pump back at our camp. Somewhere over there, far through the jungle, Katie, Oliver and the boys were slaving away, working gold on the beach. But we pressed on, heading back to the spot where we had left the canoe.

During one long rest break Wilfredo dumped everything and went off, saying he was just going to check the lie of the land for the quickest route back. He was gone so long that I fell to thinking that perhaps yet another of these intrepid jungle pathfinders had gone and lost his way. If Wilfredo had done this I would have completely lost faith in so-called trusty guides. But eventually he reappeared, so silently that I was startled to look up and see him standing there.

"Oh, Wilfredo," I laughed nervously. "You might have forgotten about me, but I knew you'd come back seeing that you left the deer and your shotgun behind."

He chuckled. "Well, I hadn't forgotten you. But you're right — I'd never go off and leave my gun behind anywhere. This gun is a dream — it's my best friend — it points by itself, I hardly have to aim. It knows just where to look."

This was far too modest, but we'd all noticed how Wilf's gun was his most treasured possession, how he fondled it and seldom let it out of his sight. Katie said it must have been the same gun Davy Crockett had used, it was so old fashioned. Wilfredo always swore it worked better when he didn't clean it.

As we chatted on our way back I asked Wilfredo if he had any family.

He paused. "I think I have," he said finally. "I ran away from home when I was seven after an argument with my mother. Then I went to live in La Paz across the border in Bolivia, so that the police couldn't drag me back home. I did dozens of jobs there — cobbler, errand boy, baker; even used to run contraband across the border into Chile, but one night we were shot at by the border guards, so I gave that up.

"I've had a girlfriend or two, and I've got a couple of kids. One of them must be almost grown up by now. I haven't kept in touch with my family, though." He shrugged. "What would I write for? They never cared for me. I'm independent here, I'm my own man and I don't owe anyone anything. It's better that way.

"I've already paid the deposit on a house in Cuzco," he went on. "A few more years of this and then I'll be moving up there to set up a baker's shop. That's something I really know how to do well."

Among other things, I thought. It was obvious that he was an expert hunter and woodsman, he knew a lot about gold prospecting and had a talent for organising his men to get them working efficiently and cheerfully. On top of this he was an accomplished amusing raconteur, an excellent cook and thoroughly good companion. Most important, he seemed to be a happy and well-adjusted self-made person, always singing to himself or joking about something.

We continued, passing a long clearing. "That's where we carved out my canoe four years ago," he said. "It was quite a job carrying it to the river, but it was worth it. That was the best cedar tree in the whole district. We did a good job too," he said proudly. "That canoe is the steadiest and best shaped one this side of Puerto Maldonado."

Wilfredo was sweating under his enormous load as we walked along, and I suggested that we should cut a long pole and sling the deer in the middle, with one of us at each end. He turned down the idea.

"No good," he grunted. "We'd get tangled up in the trees."

I dropped back a few yards to take a photograph of Wilfredo with the deer on his back. As I sighted him through the lens I saw him shift his machete to his right hand and slash out, all in one quick movement. Then he jabbed his gun barrel at the ground with his left hand. He was intent on what he was doing, and I didn't interrupt until he straightened up and turned to beckon me forward. On the ground was the twitching head of a snake. Its body was curled, quivering, round the barrel of his shotgun. I shuddered.

"I'm going to cut off the tip of its tail," he said. "If I keep that in my belt this kind of snake will never attack me."

He was quite serious. For my part I hoped that cigarette smoking was likely to be a more effective snake repellent, else I should feel I had got myself a sore throat for nothing that day.

"Is it a poisonous one?" I asked.

"No, but if it swung at you from a tree branch it could break your arm, or really hurt your head," he said. He held it up to show me. It was more than six feet long. He referred to it as a *latigo* (whip) and I've subsequently discovered that there are indeed various kinds of 'whip' snakes registered in the snake books.

Finally we came out to the beach opposite Wilfredo's farm. "I'm going to dump the deer here and go upstream to get the canoe," said Wilfredo. "Are you coming or will you wait here?"

I thought the canoe was near and went along, only to find we had almost another hour to walk before getting to the canoe. By this time I did feel very tired and didn't mind admitting it. We'd walked for eight hours at what for me was a really fast pace. So when we'd collected the deer from the river bank and crossed the river to his farm, I said I thought I'd better be getting back if Wilfredo could spare one of his men to take me.

But he wouldn't hear of it.

"No, you'd better come up to the house for a reviver. Besides you can't possibly leave without some deer and that will take them a little time to cut up."

I dragged my weary feet up to the house and Wilfredo produced three hens' eggs, a spoon and a plate. Carefully he separated the whites into the plate and started beating them up with the spoon until they were really stiff, then he added the yolks and sugar, beat it a bit more, and handed me a mugful.

"That should put a bit of life into your bones until you get some dinner," he said, pouring out another mug for himself. "I have one of these every day before breakfast, and then I don't have to eat again all day."

I nodded my approval. The simple drink tasted good.

"If you add some lemon juice or wine it's even better," he said.

Then I did get up to go. Night would fall soon, and though the brilliant moon was rising early again these days, I always felt uneasy about travelling on the river at night; it was hazardous enough in daylight. After thanking Wilfredo I said I hoped I hadn't held him up too much.

"Not at all. In fact when I'm on my own I go much slower. I imitate the bird noises, and if I hear a group of monkeys I sit down, take my shirt off and act the clown, making monkey noises, laughing and calling out. They love it, they all come and watch, then I can easily get a shot at them," he said. "We went faster today because I wanted to find a really big group of monkeys for you. The noise they make is remarkable. Pity we never ran into one. Perhaps another time."

And so I left, clutching a haunch of venison, waving as Wilfredo and his men plunged into the river for a cool swim before twilight.

Giggling and a bar of chocolate
It was too cold and wet to continue

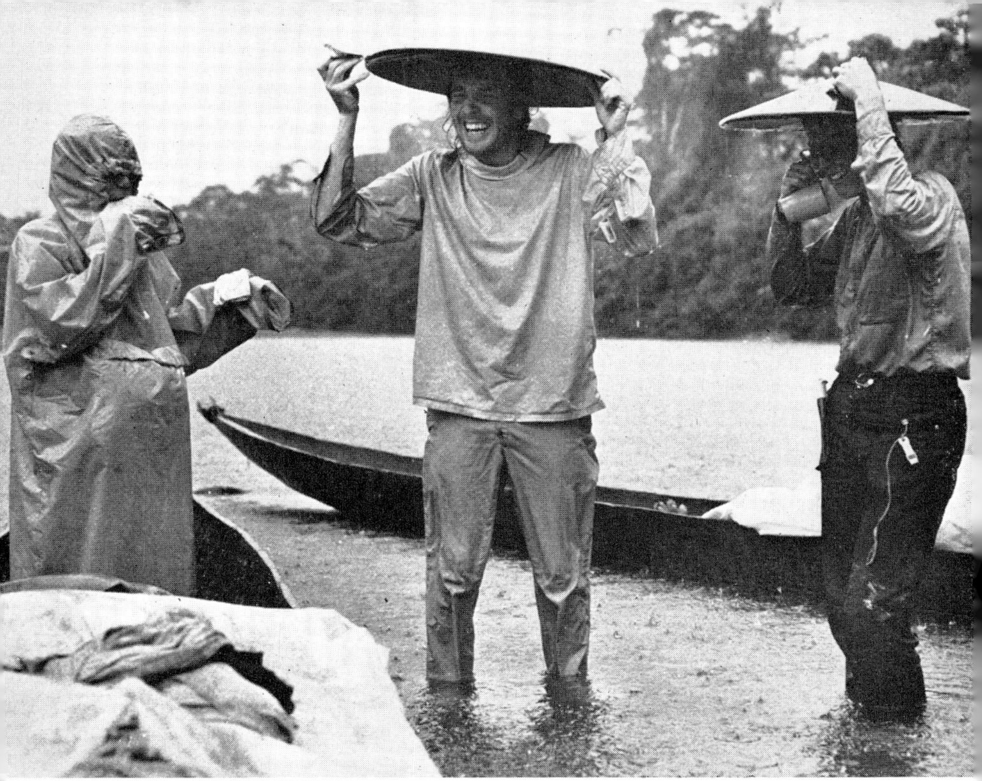

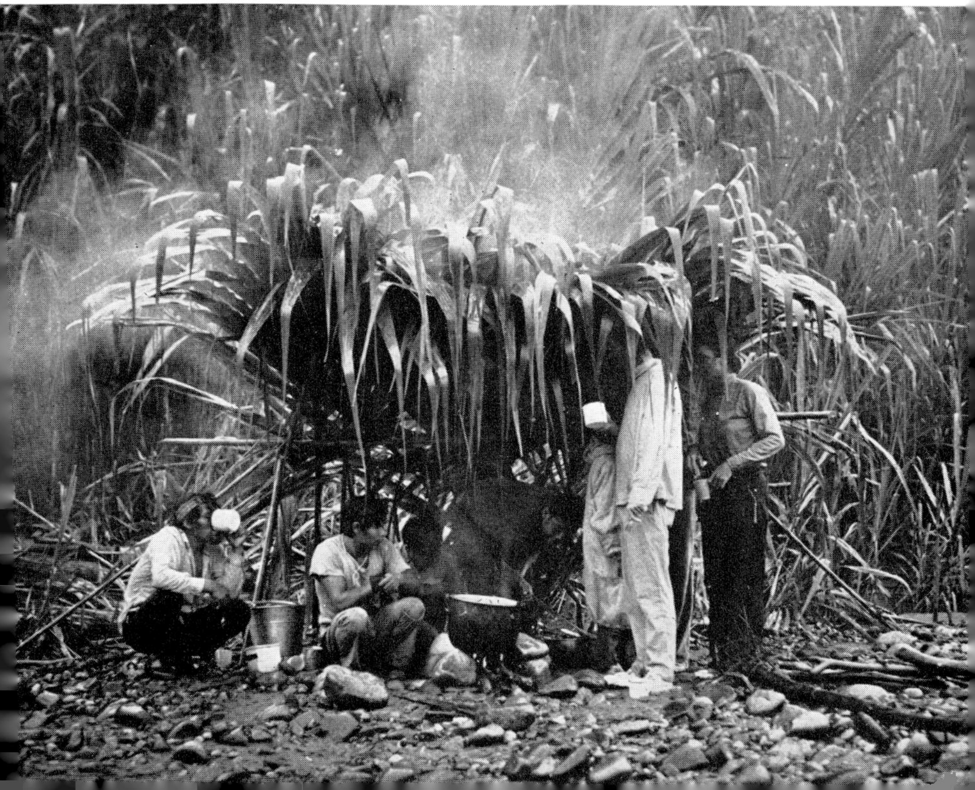

CHAPTER TWELVE

Face creams

Pills and faded rotting clothes

BEFORE LEAVING LONDON FOR our journey
I'd had great difficulty deciding which beauty products to take
with me. Make-up could obviously be left behind but I thought
plenty of skin creams were essential. The main question was
which ones I could persuade myself to dump. I am one of those
people who like to believe all the extravagant claims of cosmetic
manufacturers, so pots with hopeful names like "skin saver",
"anti-wrinkle cream" and "skin freshener" had been difficult
to resist buying. I finally decided to take along a hand cream, a
face cleanser, a skin freshener, a night cream and a moisturiser. I
was too hard on myself though, because I ended up longing for
several other favourite creams. I'd left behind a tube of face mask
—a cream for tightening up the pores—and instead at any
opportunity I rubbed the inside of the *papaya* fruit skin all over
my face. This worked perfectly as long as we had *papaya* which is
full of all sorts of healthy enzymes, but we often didn't and then I
sighed for my magic tube. The fruit of the dark-green skinned
papaya is deep orange, of the same sort of consistency as a peach
with a rich but delicate aroma.

All this smearing on of face creams was designed to fight the
battle against developing lines on my face. Hot bright sunlight

Frank hunting otter
Poling upstream
Mario and Wilfredo digging in river for gold-bearing material

provides the short cut to leathery skin and wrinkles, according to
the beauty page in all the women's magazines. And here the
sunlight was so strong at times that even with my sunglasses
on and my wide-brimmed hat pulled very far down over
my brow, I could feel my eyes screwing up against the brightness.
I should have taken a special eye cream.

Actually, the best remedy would have been to step back a
hundred years and wear a veil over my hat and face, tied with a
bow under the chin, like those elegant ladies you see in old
prints, taking the air on Brighton promenade. It would have
kept off the insects, too. As for the face cleanser, I might as well
have left it at home on the dressing table. Big city pollution
created the market for that stuff.

Hair was going to be a problem, I had decided. Katie would
have no trouble because hers is thick and very straight, and in the
event it always looked neat, tied behind in a pony tail. My hair,
though, is fine and wavy, hopeless for wearing long unless I have
a hair-dryer within easy reach. About three hours before the
plane was due to leave from Gatwick I took a deep breath and
marched into my hairdresser for a really short cut. Everyone
said that it was a great success, and I should have done it years ago.
In terms of practical grooming it proved to be ideal, because I
could wash it in the river, give it a quick rub, and five minutes
later it looked as though I'd just had it brush-dried in W.1. Our
men, though, thought I was very odd, and didn't know quite
what to make of my haircut. After a few weeks, when they'd got
to know us a bit—they were very shy of asking direct questions
that might cause offence—I got one or two hesitant enquiries
as to what was the matter with my hair. In the Peruvian jungle
there's a simple, old-fashioned rule of life: men have short hair,
and women have long hair. The jungle women, in fact, never
cut their hair, and wear plaits which grow longer and longer
until, when they are old ladies, they can sit down on them.

Both Katie and I got rather miserable at the state of our hands.

If those women in the TV soap ads. think they've got problems
with detergents, they obviously haven't had to wash up in cold
water using handfuls of sand and grit to scour the pots. Too, our
hands were soaking wet and wrinkled all day from handling
stones in the sluice trough. Our nails wore down to the quick
from this job, and soon we both looked like chronic nail-biters.
On top of this we sprouted blisters and callouses from our
stints of shovelling gravel. I just wished and wished I'd thought
to pack a tube of lanolin hand cream instead of a light cream
which hardly helped at all.

One thing which pleased me no end was the loss of pounds and
pounds of weight. No amount of slimming in London had
succeeded in turning back, even by a fraction, the accusing
dial on my bathroom scales, mainly because my will-power
collapses in the face of a promising dinner invitation. But for the
first two or three weeks of our journey neither Katie nor I had
any real appetite. We just pushed down a few mouthfuls "to
keep our strength up", as we told ourselves. I lost six or eight
pounds that I was very glad to see the back of, before my appetite
returned. Even then I didn't put them back on, because I was
getting lots of exercise. Katie was slim to begin with, and she
lost far too much weight. By the end of the trip she seemed to be
fading away, and back home she had a lot of building up to do.

Oliver was very proud that he'd gradually worked back
through the holes in an old and well-loved belt till he arrived
at one he used to use when he was fifteen. But he never had any
problems with eating—he ate masses.

We hoped the vitamin pills we were taking would tide us
over any deficiencies in our diet until we got back to civilisation.
Katie was in charge of all our pills. There were quite a few we
had to take every day, and she was the one who could be relied
on never to forget. An important pill was the tiny white anti-
malaria tablet. There isn't much malaria in Peru these days, but
there was no sense in taking chances. There were salt pills, too,

to prevent dehydration: the army had strongly recommended them. On very hot days when we were working hard we took two each, but it was a while before I could get mine to stay down at all.

Our clothes became very faded and rather rotten. After a few weeks of regular scrubbing and drying out in the baking sun, the stitching started to go at the seams. Katie landed herself with rather more than she'd bargained for one day when she offered to sew up Lucho's shirt, which looked rather the worse for wear. She called me over after many hours' work.

"Anna, what shall I do? It's like making a whole new shirt sewing this thing up. Look, I started here, and the rest of it practically fell apart in my hands." I had been just about to offer to sew up his trousers. I didn't.

"They don't have many spare things, though, do they? I haven't seen any of them with more than one extra shirt and a pair of trousers," Katie continued. "I lent Marino a sweat shirt of mine but I think he thought I had given it to him because it hasn't come back and he often wears it."

All our clothes ended up in tatters, and though we had plenty of thread with us, we didn't have the right colours to match them. As the weeks went by delicate filigrees of multicoloured stitching spread all over our shirts and trousers. I didn't notice anything odd about this until we returned to Puerto Maldonado. There I felt so conspicuous walking down the street that I had to find a market stall and buy a pair of brightly coloured jeans and a T-shirt.

CHAPTER THIRTEEN

Mazuko–Caichive

Half-way house in the jungle

I'D HEARD THAT NEAR Mazuko, on the Rio Inambari, there was an area where there were many prospectors, working *quebradas* or streams, but that their method of retrieving the gold was very different, and perhaps much better than ours. I wanted to investigate, so I took off for a week, having found a lift down to Puerto Maldonado.

From here I set off in a small plane with a Banco Minero employee, who was going to collect the Mazuko-Quincemil area gold, and it turned out to be a very exciting journey.

We more or less followed the Madre de Dios, flying at 2,500 feet, so the jungle was easy to see as it hadn't become just a solid, boring mass of green at that low altitude: every tree seemed to be a different combination of colours—yellows, mauves, reds, all types of green. The aerial view of the river itself, too, was amazing. Until you looked at it from this kind of a height it was difficult to appreciate just how much the river meandered. It went round and about in a way you never realised when you were in a boat or a canoe—no wonder it took so long to travel it! It was also easy to see where the river was changing its path, where it was going to change, and how impossible it would be to have a permanent map of it—it altered too often. We also flew every now and then in sight of the Maldonado–Mazuko road, on which a journey can apparently take anything from a normal eight hours to fifteen days, if it starts to rain, when the road surface becomes just like wet soap. As we flew further up towards Mazuko the hills began and the sight was quite beautiful: this was big country, tough and impressive, without being what you might call over-glamorous. We flew over the Rio Malinowsky, which was much narrower than I had imagined, and began to fly in among the hills just before the pilot pointed out Mazuko. It looked like two corrugated iron roofs.

"Where's the *pueblo*?" I asked.

"That's it," he said.

I had imagined Mazuko as quite a big town. After passing over a large bridge which crosses the Inambari at this point, we flew deeper into the hills and gradually started losing altitude as we followed the Rio Araza upriver and then the Marcapata to Quincemil.

No sooner had we touched down on the hard macadam airstrip than a Faucett DC-3 took off for Iberia, a settlement to the north-east near the Brazilian border. Quincemil is the second-biggest town in the Madre de Dios, and despite its unprepossessing, rundown, clap-board appearance, it is an important gathering point of a large area of mountainous wild jungle and scores of large, forgotten valleys. An area where maps don't exist—to the west it's nearly always covered in a low layer of clouds so the normal method of mapping, using aerial photography, hasn't been used till now—though new mapping methods using radar and satellite will soon be brought into use—largely due to an influx of petroleum companies who are prospecting the area for oil.

We drove off with the local Banco Minero representative to the Banco's buildings in the centre square, and the gold we had come to collect, five kilos of it, more than £5,000 worth at current prices, was packed in a dirty old holdall and taken back to the airport and off it went, straight to Puerto Maldonado.

We, meanwhile, had a wander around "town". The Bank is up in a newer part of town, the old town being further down nearer the river. It consists of one dusty narrow street with fruit and vegetable stalls and local cafés on one side and more solid buildings including the local hotel on the other side. In the old days — up till 1971 — most of the few shops and cafés used to have "*Se Compra Oro*" (Gold Bought Here) signs up but now this is illegal and the signs have come down. The street itself couldn't have been more than a few hundred yards long. Up in the new area there is a church and a hospital close by, but they do not yet have a doctor, nor, I believe, a priest. A few more houses, mostly the steep-roofed thatched jungle type, are scattered around: otherwise that's Quincemil. It proved difficult to get agreement on a figure for the inhabitants but the average guess was about six hundred. Quite a lot of lorries pass through, and on our journey to Mazuko from Quincemil, which was to take five hours, we passed about ten lorries, mostly coming from Puerto Maldonado and going to Cuzco and Lima.

We had lunch at the local hotel in Quincemil, bought some oranges, and set off for Mazuko. I preferred to travel in the back of the lorry to take photographs but it was a very bumpy ride, which killed all hope of photographs and I spent the whole time sitting on a petrol container clinging on for grim death. I suppose it would be called a normal Peruvian truck trip, but I was shaken like a jelly the whole journey without, it seemed, even a second's respite. We had one stop to pick up two gold prospectors who got a lift to where they lived. It seemed they worked for a co-operative in Cuzco and the Banco Minero had hired them to dig down in certain areas to find where the gold level was. The Banco Minero wanted to bring in tractors, but first a suitable site had to be found. They told me that although one of the holes they had dug must have been nine feet deep, they hadn't yet come across a gold-bearing seam.

We went on our way and passed through really impressive country — the same I had seen from the air — hilly jungle along the Rio Marcapata and Rio Araza, but we were high up from the valley floor and the depth of the gorge down to the river must have been at times as much as 1,500 feet and the colours were a complete contrast to those I had grown used to on the Colorado and Madre de Dios. There, the colours were warm reddish mud and the sand and stones too were warm. Here the colours were icy blue. The stones had a cold tinge and the river was more violent here, deep gorges followed by wide beaches which are obviously soon covered when the river has more water in it.

The whole region was very low in river water at that time of the year, between July and September. This apparently meant a heavy drop in gold production in this area because the *quebradas* dry out and the gold can't be worked without water.

The Banco Minero's lorry sometimes, it seemed to me, passed within inches of certain death and destruction. The precipice came right up to the edge of the road, which in any case appeared extraordinarily narrow to me. Every so often there would be a place to pass another lorry, but otherwise it was just a question of hooting and hoping.

Occasionally there would be a small group of houses. One place we stopped at was San Lorenzo which seemed to be the generally recognised half-way house where they were building a new school and had a little hut with a cross on it, and with candles burning inside.

We crossed two or three small bridges and in the dark we crossed the Inambari, and after what seemed an age, we arrived in Mazuko, which does in fact have a few more jungle huts than had appeared from the air, though not many more. It has the road going straight through and it was clear that it would not exist without it.

Next morning I went down with the Banco Minero people to the port, as the place where canoes generally pull up is generally referred to in Amazonia, basically to see off the two young men who go round buying gold—they were off upstream, to San Gaban and they carried with them a very small rucksack which seemed to have hardly anything in it. Most of their load was made up of 100,000 soles—£1,000—in crisp new small denomination notes, plus a shotgun and a small hand balance. Their journey would take about four or five days, they told me, and they added that the sum of 100,000 soles was small. Later on, in winter, between November and April, they carry at least half a million soles because they would be collecting substantial amounts of gold from the miners.

At the port of Mazuko we saw a canoe leave with a man and his wife from the altiplano town of Puno, who had been working gold in a small valley called Quebrada Nueve de Septiembre for eleven years. They had made their home there and everything they owned in the world was there. However, someone some time ago had successfully and officially claimed a concession for this area, before concessions were abolished, so these people had to pay this person one gram of gold per day as rent. This kind of thing is quite against the law and the Bank are trying to stop it— but there are plenty of people who would rather have peace and quiet than justice which might well be acquired at the cost of violence or even "eviction".

That day, too, was football day. As Mazuko doesn't have more than about fifty inhabitants it is quite difficult to get a team together. The team is in fact made up mostly of Banco people and they all turned out very smartly in yellow T-shirts and they played Santiago—a village higher up the Inambari. Once this game was over, Quebrada Dos de Mayo took the field and played against the village of Punkiri. They had all arrived in canoes down at the port, and each team, naturally, had brought along its own small quota of supporters.

The next day we started off early on our tour of the mining camps. I was accompanying Ingeniero Grimaldo Medina who, luckily for me, had brought with him a *peon* to carry my bag. I had been told to take the absolute minimum—a spare shirt and a sleeping bag, but the ciné camera weighed a bomb and so did just a very few tins of fruit and some beans. I was told that there were no gnats, so it would be much better to wear a pair of shorts as we would be crossing through the rivers and *quebradas* all the time. In the event I cheated and took two pairs of trousers as well.

The first thing was to get a canoe from the Mazuko port, which is four kilometres from Mazuko itself, down to Punkiri, a little village about fifteen minutes downstream on the Rio Inambari, at the mouth of the Rio Caichive. The Inambari proved to be a really fast-flowing river, and a completely different colour to the Colorado—blue and clear and cold looking. The canoes round here seemed enormous, and I suppose that they were used so much for transporting foodstuffs that they really needed to have a lot of space. However, neither the Caichive nor the Huepuetue, the two rivers that interested me, were navigable: in summer there wasn't enough water and in winter they were much too dangerous. So all traffic was walking traffic and the people used *acemilas*—pack animals—to carry their provisions. There were plenty of people who bought stuff in Cuzco and Mazuko and needed to get it through even as far as the Pukive and Colorado. Otherwise there were *quepiris*—boys who carry loads on their backs of up to about a hundred or more pounds, charging between 1·60 and 1·80 soles, depending on the distance, per pound weight delivered. So one of the main difficulties here was access; the enormous freight charges increased the cost of living tremendously; it can be expensive to live in the backwoods of Peru. Rice in the Caichive, for instance, was eight soles a pound while in Puerto Maldonado it was five or six soles.

After a terrible cup of coffee in Punkiri, with Grimaldo saying in disgust, "This is one hell of a dump", we set off walking; me in my smart new turquoise shorts. And did I regret it. Up the valley of the Rio Caichive were millions and millions of gnats, and I was immediately in abject misery. However, for the first three miles or so there was a kind of road in the jungle which made the walking itself easy and not unpleasant. We stopped for a time at someone's house and ate a tin of beans and sausages with some mangoes. Then off we went, walking—in my case at a half-run stumble—up the stony beaches, and crossing the Rio Caichive several times. We were headed for Quebrada Mahuay, five tributaries up on the left-hand side; in fact all the gold-bearing *quedradas* on the Caichive are on the left-hand side.

After hours of progress, crossing the knee-deep river many times, and once or twice following the path through the jungle at the river's edge, we came across our first group of gold workers—and I immediately saw that their method of working was very different from ours. For a start, the gold worked around here was flake-sized and sometimes slightly bigger, and therefore more exciting than our Colorado-type grains. Here, instead of our type of trough they had a long wooden canal set into the river bed at an angle of twelve degrees, and fourteen inches wide, narrowing down to twelve inches. Each section is about eighteen feet long, and several sections are usually joined together. In the bottom of the canal are placed fine roots of trees and plants, on top of which, in turn, are placed wooden struts of *chonta*, the hardest wood in the jungle. These sort out the gold from the stones, the gold falls between the lengths of *chonta* and is trapped by the fine white roots in the base of the canal and the roots are washed every day to collect the gold production. Primitive, effective, simple and cheap. Their method of using the stream water was fascinating in its simple ingenuity. The men shovel the material into the canal and with the twelve-degree angle the speed of the water is adequate to wash the material for the gold and at the same time strong enough to push the stones to the bottom of the canal where one of the men, the *ganchero*, shovels it away. Where the water runs away at the bottom of the canal a false stream bed is made so that the real stream doesn't dry up. From the section of the river bed that is being worked, the men were shovelling the gold-bearing material into the canal while picking out the largest stones — the size of a football and bigger — and these were eventually built into two walls at the foot of the canal so that the stream had a channel to continue flowing down. Otherwise with the heat, and small amount of water from the stream, the water would disperse over a wide area of the stream bed, and dry out. The other material already washed and being shovelled away is thrown by the *ganchero* behind the newly made channel walls.

We set off to go further up the Quebrada Mahuay and after an hour or so, having passed another small group of miners and stopped for a refresher at another miner's house, we arrived at the house of Nestor Fuentes, a miner whose operation had been recommended to me as particularly interesting. His wife was most attentive and we had tea and a delicious fried egg and bread served up in next to no time at all.

We then went to inspect their golding operation. This was the most interesting I had yet seen anywhere and it seemed to me a beautiful piece of work by any standards. It was for working the gold-bearing river banks which are steep, sometimes two or three yards high. The canal system is worked using the river water again, but in order to keep the canals at a twelve-degree angle with the first section resting on the top of the bank, the other sections are supported on stilts, so that the operation was only partially resting on the river bed. The whole set-up was neat with the false stream bed, built by the men, looking more like ancient walls than a simple construction by workers. The best use was being cleverly made of the stream with something like an irrigation channel bringing stream water to the top end of the canal.

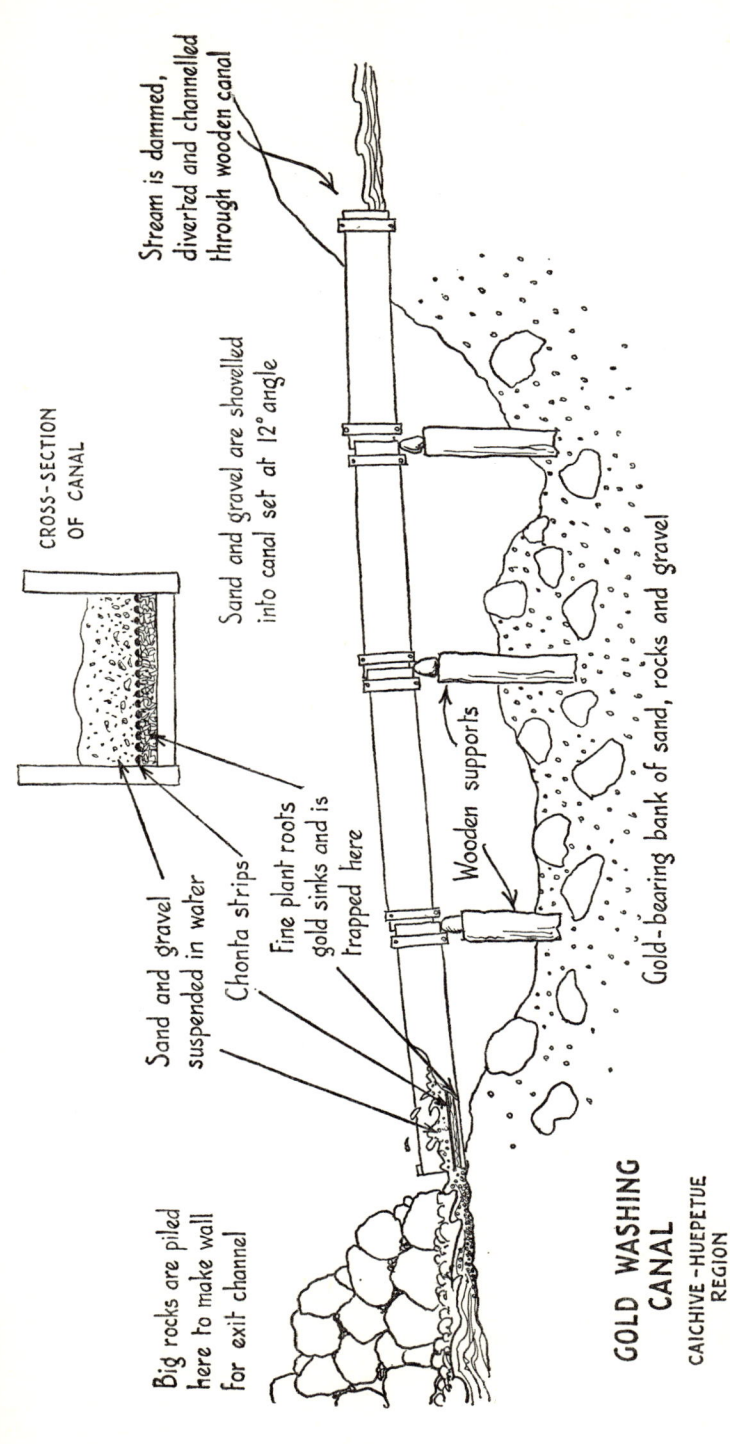

CROSS-SECTION
OF CANAL

Stream is dammed, diverted and channelled through wooden canal

Sand and gravel are shovelled into canal set at 12° angle

Sand and gravel suspended in water

Chonta strips

Fine plant roots gold sinks and is trapped here

Wooden supports

Gold-bearing bank of sand, rocks and gravel

Big rocks are piled here to make wall for exit channel

GOLD WASHING CANAL

CAICHIVE - HUEPETUE
REGION

MADRE DE DIOS

Peter Frost 1974

Just then, one of the problems was lack of water, August not really being the right time of the year to visit this area because it is so comparatively dry. From December to April is its "best" period. At this time of year the larger operations were not working because of the lack of water, and most of the big miners had gone up to Cuzco to get in more workers for the real gold-working season. So there was no point in visiting the really big operations higher up the river because no one was there, which was a pity.

Walking with Grimaldo Medina was no different from following a jungle Indian. Like them he walked at a pace which could be termed a trot or a fast shuffle. Eight hours walking at that speed, crossing through water, walking on pebbly beaches in the hot sun, with every now and then a respite going through the jungle, when they went even faster because it was cooler, proved almost enough to kill me off. When he stopped for a rest I think he counted up to twenty and then said—"Okay, let's go." Certainly when we eventually arrived at the place we were to stay the night, Moises Ortiz's thatched hut, I lay on a narrow bench and flaked out immediately and completely for a couple of hours before being woken for a delicious chicken stew. In spite of all the exercise I'd somehow lost my appetite completely and only managed to eat the chicken, leaving the rice and *uncucha*, but I filled myself up with tea and more tea which was the only thing that interested me at that moment: drinks—just gallons and gallons of liquid.

It appeared that our host's house was a favourite half-way stopover and we were not the only travellers, though we were the best-treated. A couple of Indians, a man and his wife, were travelling through, having come overland from a village higher up the Rio Colorado than our camps and, where we were to stay the night on our later journey, up to the Rio Kipoznue. There was also another miner from the Rio Pukive who was there with his group. There was, I found, a new path through to the Pukive, almost to its mouth on the Colorado, direct through the jungle—a day and a half's walk. I could if I wanted go back to our Five Bend Camp on the Colorado that way, but a day and a half's walk Indian style becomes at least three days' walk for me, and in any case there was no guide to take me.

The Indians who stayed the night had brought a *paují̇l*— bush turkey—along, so they sold it to Señor Ortíz for fifty soles, so our breakfast was secure. Rather sadly, everything around here had been over-hunted and over-fished so much that it was rare for any game to be found. So this man had to pay fifty soles for the *paují̇l* while we on the Colorado never saw anyone pay money for game. The people round here were less healthy too. All seemed very anaemic.

They didn't seem to use mosquito nets here—no mosquitoes— but we were given blankets, and we stretched out on the floor of the hut. The radio blared nightmarishly all night and the lantern was left burning to keep away the vampire bats.

Our walk the next day was back to Punkiri, but with a planned detour to visit the Banco Minero's project in the Quebrada Macho, the last *quebrada* before Punkiri, and a narrow dark little jungle stream. So, after three or four hours' hard walking to get to the entrance of this *quebrada* and then an hour's walk—fast—up it, I was nearly crying with exhaustion and misery from sore feet. We eventually arrived at this camp, which was being built in the jungle up on a bank. The Banco Minero was preparing to work this area with canals and water pumps, a very interesting scheme. Most ordinary miners wouldn't work these areas because the narrow, small *quebradas* were too difficult. Generally the ones that were being worked were wider open *quebradas* where the men could work directly, using the beach. It was only very recently that the banks and river terraces in the jungle itself were being worked for gold.

The walk back to Punkiri was luckily along a path through the jungle—because walking on the stones and through the water had practically finished me. There was a boat to take us upstream to the port near Mazuko but since telephones didn't exist round there, and since there was only one vehicle in Mazuko belonging to the Bank—which happened to be up at the offices—we had to walk from the port to Mazuko. It was only two miles, but they were long ones, after such an exhausting day.

Mazuko itself, though tiny, is the main stopping point for all lorries, and it is the place people from the neighbouring rivers and *quebradas* come to catch a lorry out to Cuzco. It is also where they bring all their provisions to be taken to the port and loaded for the rest of the journey to the mining camps out along the rivers and up the little tributaries.

However big the jungle is in terms of travel it is little more than a village from the point of view of gossip. Everyone knows everyone else's business. Next day there was some excitement when a radio message came through from Hugo Fuentes in Maldonado to say he'd heard there was a miner arriving in Mazuko who had been buying gold illegally from the other miners.

"Find him and order him to sell the gold to the Bank," Hugo said.

Everyone in the office rushed outside, but speed turned out to be unnecessary as the man in question was very much the worse for drink. He said he'd already sold his gold to an outpost of the Banco Minero back along the road. Grimaldo felt sure that he hadn't, but Mazuko had no policeman, so he had to leave it at that—for the moment. Next day the man was captured with his illicit gold in Quincemil and locked up in the local gaol.

Rio Madre de Dios

Home again!

GETTING OUT OF MAZUKO by road was risky—rain was threatening. The Banco Minero's lorry was out of commission for a few days, and I was scared of travelling on top of a passing lorry up to Quincemil as all the drivers seemed to be drunk in the local Mazuko café. When a miner who was returning down the Rio Inambari to the mouth, actually on the Madre de Dios, came in to sell his gold, I thought this ought to be a really good way back downstream. So I set the idea in motion. We all loaded up early next day into a passing lorry heading for the river, but at the port we very nearly had a false start. The condition of the miner and his *motorista* made Grimaldo Medina immediately say No—they appeared quite the worse for a solid night of beer drinking. But after a bit of an argument it turned out that the miner was, after all, more or less capable, so we loaded up and set off. Our load was ninety per cent beer—the rest being a couple of sacks of this and that.

We had hardly turned the first bend before we saw the river bank absolutely covered in about thirty wild pigs. But no one had a gun and so we just passed them by. The journey was freezing cold, raining with a chilly wind blowing. The Inambari had less wood in it than the Colorado, and was quite clear. We stopped several times. At one place, Melicio Alpaca's, Puerto Carlos, we saw one of the few well-tended farms I saw in the whole region. They had plenty of fruit trees, produced coffee and cocoa, and when we arrived they were making *fariña*. We arrived at the mouth of the Inambari just in time for me to have a mad fifteen minutes filming and photographing in beautiful light with the sun setting behind the workers on the beach. Laureano Baca, who had given me the lift, works on the Madre de Dios in the summer, and then moves high up the Inambari in the winter. In the morning he took me on down to Puerto Maldonado, and I'm pleased to say was recompensed with a supply of petrol. Later I caught a ride back up the Madre de Dios in the Bank's fast skid boat.

This was the second time I'd gone up the Madre de Dios and having seen it, in comparison with the other local rivers it is a powerful, wide, muddy reddish river that changes its mood all the time. One moment it was so flat and innocently shiny it didn't seem to be flowing at all, then the next stretch upstream would be viciously choppy and fast flowing. There were small whirlpools all over the surface, so it was easy to understand why no one in his right mind would try to swim the river – although the boat I was in skimmed across them as though they didn't exist. Often the river was over two hundred yards wide. Everywhere huge tree trunks jutted out of the water, snagged on some obstruction, and themselves acting as blockages to other branches and trunks.

The boat driver, Pajarito, who had taken us all up-river weeks ago, explained that when the river was in flood these huge trunks would be torn loose to come plunging downstream. At such times it was unsafe to travel.

Often the boatman would stand up, scanning the water's surface, reading its language. Then he'd edge the boat through a channel avoiding the débris of trees on either side.

Once in a while we would see signs of wild life. I saw a splash in the water on the far side. Something had heard the boat's engines. Once, when the deeper water channel where we had to pass was nearer to the bank than usual, we passed a whole row of brownish coloured turtles a foot or more long, sunning themselves happily on a tree trunk at the water's edge. As we came close they plopped lazily into the muddy depths. Multi-coloured butterflies darted to and fro across the river dancing their lives out while the sun shone. At one point we passed a group of perhaps two hundred bright yellow ones all fighting to eat something very special that must have been washed up by the river. Some alligators were pointed out to me, but by the time my eyes had focused in the right direction, they had submerged. Once or twice we saw herons patiently waiting for a fish at the water's edge and we saw several large high-flying flocks of small bright green parakeets.

At each bank of the river the jungle crowded over the edge. Trees seemed to be leaning over, just waiting for the next river flood to carry them away.

It was time to pull in for lunch and as the sound of the engines died we were smothered once again by the blanket of heat. Spinning along on the river with the wind fanning my face, I had forgotten how hot it really was. The insects swarmed around us, and although I'd been longing for a stop to stretch my legs, and above all give my bottom a rest—it'd become desperately tender from the hard metal seats—all I could do was gobble up the tinned sausages and dry biscuits before we set off again as rapidly as we could.

Time lost its meaning as we continued on upstream. The sun burned on, the engine droned on monotonously and banks and trees and more banks passed interminably by. By now I was used to the jungle scenery and I can't pretend it changed every five minutes.

Once at a beach where there were a few men working, Pajarito pulled in to the edge.

"This is quite a big gold operation here, run by Señor Perez. I have a message for him." We climbed the pathway to his group of huts.

A small barefoot man hurried to greet us and with a charming gold-toothed smile (gold for dentistry, by the way, has to be imported, from Switzerland usually, as unrefined Madre de Dios gold still has a trace of arsenic in it) he waved us up the steps of a hut, to sit at a long table and bench, in a welcome patch of shade. He clapped his hands and someone came running. After giving some instructions on the side, he came up to talk to us. The man had just finished the first stage of a whole new project. Instead of just working the beaches of the river, as most gold hunters do, he had cleared a patch of jungle near the camp and dug out the top soil to get at the gold from the layer of a former beach he had uncovered. I went over to look at it and saw a pit about the size of three tennis courts and maybe twelve feet deep. I forget the exact figure he gave but at the time I remember working out that Señor Perez, with his twenty men, was being very smart. He was certainly the richest man on the river.

There was an animal skin lying over the back of a chair, and when Señor Perez saw me eyeing it he got up and brought it over.

"Jaguar," he said, holding it up proudly, "if it was in decent condition people would pay 8,000 soles (£80) for it, but this one we couldn't dry out in the sun properly. We had a long period of rainy weather and it has spoiled." Sure enough there was a bald patch right in the middle of the skin. It was pretty if you like that sort of thing so I admired it politely. Eighty pounds seemed an awful lot for a skin. I found out later that it was illegal to kill this species and that this kind of price was mostly likely the going contraband rate either locally or across the border in Brazil or Bolivia. Anyone caught with one of these skins uses a standard plea to get away with it: he shot the animal in self-defence. It's difficult to argue against that.

All the huts in the clearing were built up on stilts four or five feet high, for protection against flooding. Señor Perez said that last year the river water had been lapping at the bottom steps on two occasions, and they'd been all but packed and ready to evacuate.

Our host told me how they lived. He had twenty men working for him just now, he said. They came for ninety-day stays, and he had to recruit them in the Sierra around Cuzco. Every three months he or his wife had to travel up into the Andes to hire a fresh group of workers.

"The jungle people are lazy, They're no use for this hard work," he said.

To get to Cuzco from the camp they used a different route from the one I'd taken via Puerto Maldonado. They would take their motor canoe downstream as far as the mouth of the Rio Inambari, and follow the route I had just come, down the Inambari. It was a day's journey south and west from here, following the Inambari up to the port of Mazuko. From Mazuko they'd catch one of the lorries up to Cuzco where they'd set about hiring men and at the same time load up with provisions for another three-month period.

"What do you do about fresh meat?" I asked him. He tapped a gun that was leaning up against the hut. "Either I go hunting or I send one of the men that knows the area round here. We also eat a lot of fish from the river. Quite big ones, too." Nevertheless it struck me as a tremendous undertaking to feed twenty mouths three times a day, as well as your own family, in such a remote spot. There were some hens pecking about the place and there was a large lemon tree, offering much needed shade as well as lemons, in the centre of the settlement. They grew some maize and *yuca* too, but still, it was a lot of cooking for someone — without electricity or gas.

There was a hut over to one side. It was a long structure of open framework with just plastic sheeting slung over a roof.

This was the men's sleeping quarters with each man sleeping on a bamboo platform like Katie and I had, with a mosquito net and blanket.

We had stayed much longer than we meant, so we left hurriedly to make up for lost time.

"We aren't going to make it to Camp Colorado today," said Pajarito. We had started too late and we weren't moving as fast as we should because we were too heavily loaded. "And we stopped too long back at Perez," he added.

So we drew in to a clearing, occupied by a group of huts. Muddy children, pigs and chickens were running round the place, and the inhabitants were barefoot individuals wearing shabby clothes with buttons missing and tears roughly sewn together or just left torn. I thought it was very early to stop, but by the time I had put up the light-blue tent, got some water filtered, and generally tried to organise myself it was getting dark. This was a bad area for insects, and the mosquitoes launched a major attack before I had cooked supper.

The state of these riverside camps was invariably the same. The awful poverty was distressing and at first repulsive. Refuse was discarded anywhere, and everyone went barefoot so it wasn't difficult to understand why they all had various kinds of parasites. Though these people suffer from lack of protein there was surprisingly little in the way of fresh vegetables and fruit. They hadn't got round to planting anything here although some trees and crops would certainly grow very well there as I'd seen up on the Inambari farm at Puerto Carlos and at Wilfredo Segales' place on the Colorado. Nearly all the women and many of the men and children, though naturally dark skinned, very often had a tell-tale yellowy paleness in their faces, and all had pot bellies due to parasites. They seemed, though, very clean people, washing daily in the river, and rows of clothes were always laid out to dry on the stony beaches.

A handful of rice, dried runner beans and minced beef cooked in some filtered river water made a very decent risotto-type meal, which both Pajarito and I wolfed down, and after a cup of rather disgusting hot milk made from powder, I went to sleep in my tent — Pajarito was given a blanket and mosquito net up in one of the huts.

When dawn came, we got up quickly, packed and set off. I still had a sore bottom from the previous day's hard ride so today I abandoned caution and used my life-jacket as a cushion.

I had seen very little river traffic in all the time we'd been travelling. Only twice had we passed other craft going down-stream, motor-powered canoes chugging away, with their passengers waving gaily as we passed by. This morning, though, I saw someone else. There, coming towards us at a snail's pace were eight or nine men standing on tree trunks in the middle of the river.

"Woodcutters. They lash the tree trunks together and steer the raft with that huge oar at the back there. The bamboo thing in the centre is their shelter. At night they pull in to the side and sleep inside. They're taking the logs to Maldonado, and they should arrive there in five or six days," Pajarito explained. Five or six days! And we had just covered that distance in a few hours. So that was how long it took to float down.

Another time we overtook a canoe so loaded it looked as though it were ready to sink at any moment. Our boats in fact had to give it a wide berth, so as not to swamp it with our wash.

They were, it turned out, professional hunters, who disappear into the jungle for a month or six weeks, and then return to Maldonado to sell the skins. Pajarito reeled off the names of all the kinds of animals they shot.

There is something mesmerising about a long river trip like this one. I was short on sleep from the previous rather uncomfortable night, and I caught myself nodding off several times, with my chin resting heavily in my palms.

Suddenly we had arrived. Back at Camp Colorado. Later I caught a canoe up to our own camp, to Katie and Oliver.

Home again!

Finally, they arrive

Tarantula
Packing up

VICTOR VARGAS WAS BACK in the area. He had brought in some badly needed articles plus mail while I'd been away. He had promised the others to come up again as soon as I was back which was just as well as it turned out. Marino came up to me at camp that first morning. I thought he had already set off to help put out the tubing for work.

"Oh, did you forget something, Marino?" I asked.

"No, *señorita*. I came back here because I wanted to speak to you," he replied. "You see, I want to leave immediately. If necessary I will build myself a raft and go now."

"But I thought you were enjoying yourself. What's wrong?"

Marino then blurted out, "While you were away Lucho and I had some very bad words together. I know you told us that you wanted us to stay on with you because we were friends and worked happily together. Well that isn't so any more, and I don't think it would be right for me to stay on. We aren't speaking, and it would make a bad atmosphere. So I want to leave right now."

I had to tell him that if he didn't want to stay on I couldn't force him to, but he would do much better to wait till Victor Vargas came so that he could get a lift downriver. And so he left us that afternoon. We would miss his laughing cheerful manner, and his good cooking. It was sad to lose him so abruptly, but we could all now feel a heavy atmosphere when he and Lucho were even in sight of one another, so it was clear that one of them had to go.

We had still not been able to get hold of any wooden pans for ourselves and after some persuasion Wilfredo had agreed to make them for us, with Lucho's help.

We spent the day at his place while they both started working on two pieces of tree trunk that Wilfredo had been storing for eighteen months to season the wood, he said. They gouged out wood from the large blocks, going deeper in the centre. They axed off surplus wood from the back and when they'd roughly hewn out a shape they marked the centre and with a piece of string and a sharp tool they drew the circle—then they chopped away with smaller neater movements turning the block as they worked to produce an even surface.

After several hours' watching I got bored and went off with Juan to look for turtle eggs, as we were back on the bread line again having not been hunting recently. We found hundreds of eggs, after several hours of searching and when we got back we found we'd missed lunch, *and* the best part of a bottle of wine.

"Wow, where did that come from?" I asked Katie.

"Wilfredo just produced it. There was some story of a motor canoe stopping yesterday but I didn't quite understand all of it," she said. Wilfredo produced another bottle of wine and we all sat around in the heat of the early afternoon drinking the wine in a relaxed manner. Around mid-afternoon, Wilfredo suddenly said:

"Could I go up with Juan in the motor canoe to check the men are working all right, Señorita Anna?"

"Of course," I said. "We'll come too." And Juan, Wilfredo, Katie and I set off upstream after polishing off the remains of another wine bottle. As we reached the beach where his men were working, Wilfredo called out a greeting, waved to his men, and signalled Juan to keep going.

"I just want to visit some friends up there and round the corner,' he said.

The heat on top of the wine made everything rather hazy. Anything Wilf wanted was okay with us: if we could do him a favour for once, all to the good. A couple of times the river was so shallow the canoe had to be pushed. I jumped out to help over the difficult bits, soaking myself completely, but it didn't seem to matter. We all chattered merrily and though we found out afterwards that the journey had taken one and a half hours it seemed at the time more like twenty minutes.

Wilfredo's friends lived at the mouth of the Rio Pukive, which we'd been told was a day's journey away. So much for the fables we heard round here. No one ever knew how far away anything was in distance or time. These people had a large group of huts and a shop, quite well stocked, but as usual we didn't have any money, so we couldn't buy anything. They offered us glasses of wine and proposed many toasts. The whole district was seething with excitement. A consignment of wine had arrived from Puerto Maldonado. Wilfredo did not say what it was he had to see his friends so urgently about, but the white sack he brought away with him looked suspiciously as though it might contain many many more wine bottles.

We got back very merry and gay to the relief of Oliver, who had become worried by our delay. Then the whole camp settled down to a long, heavy session of drinking and toasting in the Peruvian custom. The method is simple and effective. Anyone at any time who feels like taking a drink raises his glass and shouts, "Salud!" whereupon all those present empty their glasses in one gulp. When you're drinking like that there has to be a lot of booze around, and that night there was.

"Juan, how will we ever get back to camp?" I said, peering at the river. "It's nearly dark."

"S'all right, *señorita*, look at the moon!" he cried, waving his arm expansively at the sky. "It's like driving in daylight when it's as bright as that, and later on it'll be even brighter."

And so we stayed on, stayed for supper and more wine, more drinking, and toasting, until I could stand it no longer and left the table, picking my way unsteadily between the huts to the pleasant solitude of the river bank. Presently Katie loomed through the darkness and plonked heavily down beside where I was sitting under a banana tree.

"Katie," I mumbled, "my head hurts, Let's go home."

She felt the same way and, supporting each other we set off for the canoe. Propping each other up, Katie and I slithered and sprawled down the muddy bank, giggling helplessly and asking each other's help.

"Olly," I called, for the sixth time. "Please will you look after my camera?" For the sixth time he assured me that he was guarding it well. Somehow he had contrived to down fewer "Saluds" than we had; either that or he was made of sterner stuff than us. Anyway, he was definitely in control of the situation. As for Juan, he came wobbling down the bank, walked straight into the river and fell over, disappearing entirely beneath the water. He emerged calmly, with water streaming off him as though nothing had happened and drove us back to camp confidently as though it were broad daylight. Somehow Wilfredo was with us, wrapped in a blanket and very much the worse for wear. When we arrived Lucho and Juan carried him up to the orange provisions tent, and left him inside to sleep it off.

It was about ten o'clock when we got back—late for us—and we all went to bed immediately. Somehow wine didn't have its usual magic knockout effect that night, though, because at about three, Katie, Lucho, Juan and I were up attacking a midnight feast. Luckily I had made one of my marathon three-day rice puddings that morning. Suddenly there was a yell of fright from the orange tent, and Wilfredo came scrambling through the entrance in a panic. He relaxed when he saw us sitting round the fire and laughing, but it had been quite a shock for Wilfredo, waking up like that in a strange tent. The last thing he remembered, he told us, was strolling down to the river, planning to see us off.

He was still pretty plastered. "I want to go home," he said.

"Can't be done, Wilfredo," we told him. The moon was down by now, and Juan luckily refused to budge the canoe in any case.

"But I can't sleep in that prison with those dark walls," he complained. But Lucho managed to hustle him back to the tent, and peace reigned in the camp for a few more hours.

At six Wilfredo was up again, dragging Juan out of bed and demanding to be taken home. Katie and I got up much later, giggling sheepishly, feeling awful, and taking twice as long as usual to prepare breakfast and do the washing up. We discovered our clothes were covered in mud. "Goodness, I've even got it caked in my belt buckle," said Katie in amazement. We stared dumbly, wondering where it had come from. Only then did we remember our undignified tumble down the river bank the previous night.

The day before we had been presented with a dozen hen's eggs by a lady at the shop, but not even this rare treat could tempt me to have any breakfast. Only Oliver was looking bright that morning.

Around mid-morning Juan reappeared. Wilfredo was still sitting there in the canoe, as though he had never intended to leave us at all. The pair of them were pie-eyed once more. Wilfredo was clutching a bottle of wine, offering it around to all takers.

But nothing could induce Katie or I to have any more, so Juan took Wilfredo back home again. He returned at nightfall very quiet and with a sore head, saying Wilfredo had sent his men off to get another two dozen bottles of wine. As a result of last night's binge, we had missed our usual radio programme and it wasn't until Victor appeared at midday that we discovered Nick and Igor, my brother and nephew, were at last on their way, having arrived in Puerto Maldonado . . . We waited expectantly for two days, rooted to the camp, and still they did not show up. What could have gone wrong now?

It was a relief when Wilfredo came paddling down to ask if we could take our canoe upstream to look for two of his men who had disappeared with his radio and canoe. We all set off to help wondering how anyone could get lost on this river. We had been avoiding Wilfredo's place for the last couple of days, knowing his camp would still be swimming in wine. Once was more than enough for us.

His men were an hour upstream, very much the worse for wear. But since Wilfredo had been in much the same condition himself he only scolded them half heartedly, playfully pretending to drag them by the scruffs of their necks to our canoe, tying the canoe his men had taken to our bows and holding the stern tightly. When we returned to our camp, there was my brother, with ten-year-old Igor, waving madly.

"We thought you'd never arrive so we decided we'd go off and do something. We've been waiting for two days," I called as we came in to the river's edge.

"The boat broke down and we had to come up slowly," said Nick. "But we've had a really good time on the way. We've been fed on turtle's eggs and venison and we've stayed at some great places along the Madre de Dios."

"By the way," he added, "we've brought you a few provisions you asked for. They're over there in that suitcase." And it was full of apples, bread, cheese, ham, even some rather soft butter, lemons, oranges, and plenty of tins of fruit. We wasted no further time on conversation, and began eating as though we'd never seen food before, stuffing ourselves with bread, butter and jam.

"How long can you stay?" I asked my brother, when I had eaten enough.

"Well, it rather depends on your movements. A week or so," he said.

"We have been longing to get further upstream," I told him. "We talked to Wilfredo about it and he says up in the Rio Kipoznue we can pan for what they call *charpas*—little nuggets—but it's about eight days poling up stream from here. The motor won't take us up very far. It gets too shallow." I went to fetch the map.

"Have you asked Wilfredo if he wants to go?" asked Nick. We had not, but we thought Wilfredo would come and show us the place. It was agreed that we should go and ask him right away, and Wilfredo said he would leave as soon as we were ready.

And so it was that we found ourselves packing up our camp. Nick decided that he and Igor would just have to find the time to come with us. Nick was due back to edit the *Peruvian Times*, and Igor was flying to England in three weeks' time to go to school. But this was the journey of a lifetime. We were off, at last, to the end of the rainbow, to search for lumps of gold and we had a better guide than anyone could have hoped for in the shape of Wilfredo, who was all keen to leave as soon as possible.

Lucho seemed keen, too, to accompany us. Juan was to be given the job of taking us as far up the river as possible in the motor canoe. We would attach Wilfredo's canoe to the bows of our motor canoe, as his had no motor. Then we'd pole on up. On his way downstream Juan would pick up the rest of our equipment—more than half of it—which we had had to dump because we couldn't possibly fit it all into Wilfredo's canoe. Juan would leave our things at Camp Colorado for us to collect later. He himself would return to Puerto Maldonado.

We were heading, on Wilfredo's advice, for the Rio Kipoznue, a tributary of the Rio Colorado, that had been the scene of a minor local gold rush about eight years previously. Wilfredo said about sixty men had worked up there, and most of them had stayed for several months until they had two or three kilos of gold each. Then they'd all come home again.

"Mind you, there were really some bad fights up there about whose patch was whose," he went on. "The Kipoznue is a dangerous little river. Why, at least three canoefuls of beer have been lost there just in the time I've known it."

That remark set the seal on our experiences of the past few days. Only now did we realise just how important alcohol was to the people up here. When a big consignment arrived, everything stopped. The routine death of a man by violence or drowning was nothing compared to the tragic loss of an entire canoeload of beer. A lot of gold, it appeared, was getting converted directly into beer or wine, hardly even being cashed into hard currency.

Our packing was a very tiring affair. Since we didn't know what to expect up in the Kipoznue we had no idea what to take and what to leave behind. Katie and I realised we would have now to sleep on the ground as the others had done. Since we would be travelling for weeks, and making a new camp every evening, there would be no time for building beds, nor for digging in hammock stakes. We decided to leave behind the big orange tent that we had been using for storage. This was a luxury and with our waterproof packs and numerous plastic bags we should be all right without it. Wilfredo said the trip was tough, that he personally would make sure that Igor was all right. Igor couldn't decide whether it was more fun missing school or going on a real goldhunt into the depths of the unknown jungle. Nick decided that the *Peruvian Times* had been going for sixty years and could manage for a couple of extra weeks without him. "I'll just say we were surrounded by Indians and couldn't get away," he said.

All the gear was divided up, and everyone came up with interminable problems and queries:

"Anna, what are we going to do about the paraffin? Does it

Rio Kipoznue Camp — smoking monkey meat
Mario's big catch

go?" — "Katie, what are you doing about the ciné cameras?" — "Who's going to be in charge of the soap?" — "What food shall we keep out for the journey?" And so on, well into the night. I started packing up the food, trying to fit it into two big waterproof bags, transferring sacks only half or a quarter full into other plastic bags to save space.

Once I stretched out my hand, groping vaguely for a partly filled rice bag and had nearly closed my fingers on it when I saw something that made me recoil in horror. I had almost picked up a brown furry monster, whose body alone was easily the size of the palm of my hand and quite horrifying to look at. It was a tarantula; I'd seen enough films to know that without having to be told. I hope it's to my credit that I just jumped back quickly and in my best don't-show-the-men-you're-frightened voice I managed to choke out:

"There's the biggest spider I've ever seen in my life, in the rice." Oliver, looking like Douglas Fairbanks, drew his machete and leapt forward in a flash and quickly chopped the beast in half.

Back in England I checked up to find out what would have happened if the spider had bitten me. But there are a lot of different kinds of tarantulas and, while I probably wouldn't have died, I could have had an excruciating pain for a few days.

In the morning we had planned for an early start, but we just couldn't manage it. There was still all our gold-sluicing equipment to dismantle and load up. We weren't taking this with us as we would be panning only, and just needed pans and shovels. Wilfredo and Lucho had succeeded in finishing one pan for us the other day, before the wine arrived and the rot set in. But Wilfredo said he had plenty of extra pans of his own to spare, and could also lend us a special small one for Igor. Igor and Wilf had immediately got on together like a house on fire. Igor's small pan was about fourteen inches across, while ours were some

Panning in Bandera creek

twenty-four inches across, half an inch thick and about four inches in depth at the centre of the cone. We had watched Wilfredo and Lucho gouge and chip away at a pan for us with incredible dexterity, giving it a final finish as smooth as though it had been planed away, though the only tools they had to work with were an adze and an axe. This pan meant a lot to us, much more than if we had just bought it on a street corner in Puerto Maldonado. When we eventually left for Lima I took it with me, and brought it back to England.

Even though we had cut down our gear by half, we still managed to fill our canoe almost completely. How we were ever to get it all into Wilfredo's canoe would be a mystery till the transfer moment actually arrived. We set off without breakfast, only keeping out some tins of fruit and a box of soda biscuits we had. It was like moving house. Last-minute things like plates, mugs and spoons were thrown into an empty bucket; toilet paper, carefully preserved in a plastic bag, was stuffed into some crack; spare ropes were thrown on top of everything; and plastic sheeting was spread out over the whole load. It was a rather grey, drizzly morning, not the sort of encouraging weather we wanted for a long canoe trip.

Just as we were about to leave Victor arrived, having come specially to see us off. He was looking much happier, as he said his wife had recently arrived up at the camp. We were able to return his radio, and his table, both of which we had used consistently. We'd all agreed that on a journey such as this one should set oneself up properly with canvas stools and camping table at the very least. Our lovely bamboo table made by Marino was finally and sadly abandoned. It had made a lot of difference being able to sit at a table rather than having to stand around holding a burning hot plate as we ate. The table had helped, too, for cooking and dishing out. One of the small but tiresome problems we had had to put up with was the sand from the beach which got into everything including the food. But by being able to prepare food on a table, we could reduce those gritty bits in the rice pudding to an almost acceptable minimum.

Our canoe pulled out into midstream and with a lot of hand-waving and Good Lucks and *Buen Viaje* — Bon Voyage — from the small group of Victor and his men on the beach, we moved off slowly rounding the bend towards Wilfredo's place.

As usual we were much later than we had planned.

"You were supposed to arrive two hours ago," said Wilfredo reproachfully. "If we leave this late every day we'll never get there. We must start off at four-thirty each morning, and go all day. We will only eat at night and if there is anything left we will have it cold in the morning. If not we leave without breakfast." Wilfredo had definitely taken charge and we must all have looked a bit shamefaced. We just hadn't realised how long our packing up would take, I explained feebly.

"I've decided to bring one of my workers with me. We're going to need some help later on. Besides I want a helper to work with me when we get to where the *charpas* are." Wilfredo and his worker, a young boy called Mario, had three great sackfuls of food, and a large bag of salt. We couldn't imagine why they needed so much, and felt that we hadn't done so badly in saving weight after all, if those two were bringing so much. Little did we know then how glad we were to be that they hadn't cut down as drastically as we had.

While he packed our canoe to make the load sit better and lashed his own canoe to the bows, we stuffed ourselves as unobtrusively as possible with a makeshift breakfast of tinned fruit and dry biscuits. Although we hadn't arrived until well after 8.00 a.m. we'd been working hard since dawn.

Then it was time to set off, and we clambered into the heavily laden canoe. As we pushed out into the current Wilfredo tossed back a final word to his men.

"Angelitos," he called, laughing gaily, "there'll be some beer for you if you work hard while I'm gone."

Waving cheerfully, and in great spirits, the nine of us started slowly upstream. We were all wondering how long it would take us to get there. And since we were going right off the map, we also wondered where exactly *there* was. Only Wilfredo knew that.

CHAPTER SIXTEEN

Upriver and off the map

Floating amateur gypsies

THE LITTLE OUTBOARD MOTOR whined and strained, trying to push two canoes, nine people and a huge mound of gear up the fast-running river. We seemed to be almost stationary, just barely holding our own against the current. The river bank crept by terribly slowly, just enough to convince us that we were actually moving upstream.

That day Juan was going to drive us up the Colorado as far as the river would allow, until the shallows and the obstructions were permanently fouling the propeller and we had to resort to poling. We were so low in the water in the Bank's large canoe that I, for one, still had the gravest doubts as to how we were going to fit everything into Wilfredo's much smaller canoe.

Slowly as we seemed to be moving, we managed to overtake another canoe that was poling upstream. There were three young men aboard. Wilfredo knew them and waved, exchanging jokes as we passed. They looked unusually neat and efficient, with their load all carefully stacked and covered, keenly poling upstream at a fast pace. Wilfredo told us that they were going out to look for gold for the first time, heading up the Rio Huasoroco, a tributary of the Colorado, and planning to hunt skins—and look for valuable woods as a sideline. They would be away for at least three months.

What impression did they have of us, I wondered? We were a strange-looking group, like floating amateur gypsies. Pots and pans cluttered the bottom of the canoe and there were bags of fruit stuffed into every free corner. Spread out over the cargo was a multi-coloured patchwork of damp clothes, steaming gently as they dried out in the hot sun. Everyone was hunched or sprawled in his own little world, unable to converse over the noise of the engine. Katie was catching up on a bit of sleep, curled up in the back. Igor in his peaked Hemingway fishing cap looked alert and ready for anything. Nick had slipped quickly and comfortably into acting his new role. He already looked like an old-time goldminer, wearing a battered straw hat and a five-day beard. All he needed to complete the image was a bottle of whisky in one hand. No chance up here.

All of a sudden Wilfredo's canoe—with Wilfredo in it—and ours parted company, and Wilfredo paddled energetically on ahead with Mario, his helper. A little further on we arrived at the reason for this manœuvre: rapids. Everyone had suddenly to wake up and pile over the side to lighten the load and help shove the heavy canoe up the fast-running shallows which in addition were also blocked by tree trunks and branches. Wilfredo and Mario had easily pushed their empty canoe over the difficult stretch, and they came back to help haul ours over a submerged trunk. Sometimes the boys were waist deep, trying to get beneath the rear end of the canoe to swing it up and over the obstacle. Then there was a shout as the canoe slid forward and everyone gave a final push together.

"That's the first of many," said Wilfredo, grinning at us as we sprawled exhausted in the canoe. "There are eight days of this, perhaps a little less if we make good time. But the rapids get worse, and there are places where we'll even have to unload the canoe and carry the stuff up the bank."

Further on when we came to another shallow section I was pushing Wilfredo's canoe, and Mario was tugging at the bows. He seemed uneasy, and I asked him what was the matter.

"*Señorita*," he said, "this is a very bad place for *rayas*—stingrays—and I'm at the front." I didn't offer to change places.

As we hitched up Wilfredo's canoe again and started up the little motor, a canoe appeared round the bend, heading downstream. It was the Sumalave brothers—our hosts from the previous week, the people who had invited Katie and I to share a bottle of wine with them that tipsy day with Wilfredo. The brother had broken his leg a month earlier in a drunken fall, and the bone had failed to mend. Now he couldn't stand the pain any longer, and his family was taking him to Cuzco, via Puerto Maldonado, for medical treatment. Later Wilfredo was to tell us that this injury was only one of their worries. Both the brothers had got into a big local dispute, some trouble over a woman. Their rival had apparently complained about the family to a newspaper in Cuzco, which had printed his story accusing them of numerous crimes, including maltreatment and cheating of their workers, sexual deviations (not specified) with unwilling women, and actual murder of one of their employees. Libel laws in Peru are less strict than the ones we have in England.

As these people were friends and, by the long-distance standards of the river, neighbours of Wilfredo's, we stopped while he chatted with them. We gave them some Panadol tablets to relieve the man's pain on the six-day journey to Cuzco that lay ahead of him.

Soon after we left these people our motor died on us. Probably the strain was just too much for it. Juan tinkered with it for an hour without success as we baked in the heat and we finally decided to abandon the boat and make the transfer to Wilfredo's canoe even earlier than anticipated. This took some time and some delicate manoeuvring of all the bits and pieces by Wilfredo. When the canoe was fully loaded we all looked at it doubtfully, even Wilfredo. It was stuffed full and riding very low in the water, and there wasn't even anyone sitting in it yet. We all settled in cautiously, and Wilfredo, Mario and Lucho began poling. We all waved goodbye to Juan as he turned away to float downstream.

"There's only room for three of us to pole," said Wilfredo. "Whenever we can we'll drop you lot off to walk along the beach. That way we should make good progress, but if you all stay aboard it's going to be very slow going.

"Now," he said, preparing to push off, "don't anyone move unless I tell you to."

We all thought this suggestion that we should walk on the beaches was a splendid idea. Nick announced in midstream that we had two knucklesworth of his middle finger of free board in the middle, and I thought we had about three inches where I was sitting in the stern. The slightest ripple threatened to swamp us.

We made our way precariously upstream, plodding along many stony beaches, jumping in and out of the boat as Wilfredo and the boys crossed and recrossed the river, always seeking the channel where the current was weakest. Finally we were dumped on the bank, and Wilfredo pointed to a path leading into the jungle.

"Follow that, and keep going until you come to an Indian village on the river. Wait for us there.

"There's a big bend in the river just here," he explained. This turned out to be a forty-five minute walk. Often we stopped, looking at the ground anxiously, for in many places the trail seemed to just peter out. Once we were relieved to come across a group of Indian women feeding their children, doing some washing and looking rather coy. We asked them the way in Spanish, but of course none of them could understand us. But they smiled, waved their hands and pointed with their heads to indicate the direction we should follow. At length we came to the village, a group of about twenty huts with a bare earth clearing in the centre—the indispensable football pitch. Except for a few women and children the village seemed deserted.

We strolled over to the river's edge, waiting for the canoe to appear. Eventually an Indian came up to shake hands, greeting us in Spanish. We recognised him—it was the spokesman for the group that had visited our camp a few weeks earlier on their fruitless quest for beer. Our friend promptly took up where he had left off at our previous meeting. Would we like to buy *tigrillo* skins? He showed us some of these furs, and they were indeed very beautiful, but we persisted in declining his offer. This was not the time to be loading ourselves up with surplus weight even had we wanted to. He gave us bananas, which we accepted gratefully. We were hungry by then.

Half an hour later the others arrived. It was now late afternoon, and we made arrangements to stay at the village. Wilfredo said he was planning to try and rent another canoe here, as one was just not enough for our party. He went off with Nick and Igor to talk to the village schoolmaster about it. In these isolated little communities, the teacher is often a respected man and a good intermediary when one is trying to get something done.

Katie and Lucho spread out our bedding and mosquito nets under an open-sided thatched roof with a raised platform beneath, which the villagers had offered us as lodgings for the night. Oliver set up the water filter, and I got down to making some supper.

Before long Nick and Wilfredo reappeared looking smug. "We've got a canoe *and* an Indian," said Wilfredo. "I thought it would be a good idea to take an extra man with us, because he can hunt while we all go off looking for the gold."

It was soon dark and we then got our first taste, on this journey, of danger. As we stood around eating supper some Indian women nearby started jabbering excitedly. There was a bit of a commotion and Wilfredo jumped down from the platform to see what it was all about.

"Keep away!" shouted Wilfredo suddenly. "Snake! Snake!" Half a dozen torchlights zeroed on a rapidly-moving brown snake which made a beeline for our platform. After a vast brouhaha it was beaten to death with long poles and gingerly carried away into the night.

"Any of us could have walked straight on to it," said Oliver, and he was right. No one was at all happy about this as the snake, about five feet long, was a fer-de-lance—*jergón*—which is highly venomous.

"This is the time of the year for them," said Wilfredo conversationally, but I could see he didn't like the incident at all.

Wilfredo started up an interminable monologue concerning his eating and sleeping habits.

"While I'm on a journey like this I rarely eat in the daytime. There isn't time to stop, you see. So I reckon to eat several meals during the night. I don't need much sleep; between meals I just listen to the radio and play cards. Tomorrow we don't have to get up too early, because we've got to wait for our Indian and his canoe. But most mornings we should be off by four-thirty, otherwise we'll never make it."

Wilfredo's colourful and expansive description of his life-style on long voyages had been very entertaining, but we allowed for a bit of exaggeration, and I don't think anyone took it too seriously. But that night we discovered that he wasn't kidding us at all. We had all crawled into our sleeping bags soon after supper. Lucho and Wilfredo stayed up, conducting a noisy and impassioned game of cards in which, Nick assured me, several grams of gold were at stake. Beside them sat the radio, blaring away full blast.

It was difficult to sleep in the midst of all this racket after the peaceful nights we had known for so many weeks. At eleven o'clock I heard the pair interrupt their game to fix themselves a meal of spaghetti and tuna fish. Later, at three in the morning, I woke to the sound of Wilfredo shouting at Mario to get up and put the kettle on. The radio was still going. As Wilfredo had said, he didn't sleep much.

We rose wearily at five—the stars were still out—to eat a breakfast of porridge and dried apple. The schoolteacher later appeared, and insisted on showing us the village school house before we left. It was a little thatched hut with a dozen small desks on the earth floor, walls of bamboo, and a map of Peru hanging at the front of the class.

When the second canoe arrived Wilfredo sorted us all out into groups, putting Katie, Igor, Lucho and Oliver into his canoe, while Mario, Nick, the new Indian that we'd nicknamed Frank— real Spanish name Francisco—and myself went into the other canoe. I soon realised that Wilfredo hadn't been boasting idly when he had extolled the stability and general excellence of his own dug-out. I had difficulty just standing up in this one. I poled cautiously as we set off upstream. Soon we pulled into the bank at a good spot for cutting bamboo.

"The cane down here is much better quality than the stuff you find higher up," said Wilfredo. "We'll cut a good supply to take with us."

Both canoes set off with a big load of *tangos*, as Nick called them, laid out on top of the cargo. *Tanganear* is the Spanish for the verb "to punt", so while Wilfredo and Lucho used a lovely rolling Spanish participle—*tanganeando*—to describe what we were doing as we moved slowly upstream, Katie and I decided that we were just "tangoing". It was easier on the tongue.

Not far upstream my tangoing came to an abrupt and undig- nified halt. The canoe struck a submerged log, throwing me off balance, and I somersaulted into the river, glasses, hat, belt, penknife and all. In passing I bumped domino-like into Nick and he came splashing in beside me. We both surfaced together, laughing so hard it was difficult to get back into the canoe. It wasn't a place I would normally have chosen to bathe in as it was very muddy, but it was deep enough for us not to bang our heads on the bottom. Wilfredo had taught us well.

"If you lose your balance," he'd said, "don't try to hold it whatever you do, or you'll turn the whole boat over. Just jump in." To our credit we had got it right first time and we soon became experts. The occasion was also notable as being the first time we saw Frank, with a normally very deadpan Indian face, smile.

Oliver in the other boat seemed already to be pretty stylish with his tango, but we noticed more than once that there was no rhythm in their poling. Mario, Nick and I were keeping time, and finding things much easier as we didn't clash with our poles, but the others seemed to be having a bit of a struggle. We discovered the reason for this later. Wilfredo quietly mentioned to us that Oliver was doing all the work getting that canoe upstream, because Lucho was only pretending to push, and making hardly any effort at all. Wilfredo said he had spoken to him about it several times without it having any effect. He obviously hadn't expected to have to work so hard.

Oliver was looking very sporty that day, clad only in shorts and a T-shirt. It was relatively cool and there were few mosquitoes in this area, but the sun was strong and Oliver's legs got very sunburnt. Katie too was happy with a shortsleeved T-shirt, but I stuck conservatively to long sleeves. Katie had a very fluid and ladylike style with her tango, and she seemed able to keep going for hours without letting up. Later, though, I talked with her, and we both admitted sheepishly that after all it *was* possible to seem to be tangoing without actually making all that much effort, when one was too tired to keep pushing. With all Wilfredo's exhortations about starting early and keeping going, neither of us had the nerve to demand a rest break.

We passed only one group of goldminers and they proved to be the last we were to see. They were friends of Wilfredo's from

Cuzco. Isaac Margol, by name, and Wilfredo was godfather to one of their children. These people had come to their camp by walking for four days along a difficult jungle path and then making a one-day canoe trip. They were accustomed to spending six months of every year in this isolated place. The woman had just had a baby at the camp a week previously, so she must have been over eight months pregnant when she made that four-day jungle walk. She had two young children with her too.

We continued on our tiring journey pushing hard while Wilfredo sung Peruvian *huaynos* and folk songs, and yelled out ribald wisecracks every time someone overbalanced and fell in — which was often. There were times when we'd have to jump out and push the canoe over shallow rapids. It wasn't difficult, but cumulatively it was excessively tiring.

At about four o'clock Wilfredo called a halt saying it was time to set up camp before nightfall. There was quite some discussion as to whether we should have the plastic sheeting up. The sky looked absolutely clear, but we all wanted it up just in case. Lucho, though, obviously couldn't be bothered, and only grudgingly helped Oliver after some stern words from Wilfredo. Wilfredo, Frank, Mario and Lucho decided to sleep out on the open beach near the fire. Almost before we'd unloaded both Frank and Wilfredo had gone off with the fishing lines without even bothering to unpack their sleeping bundles. "We don't need mosquito nets up here," said Wilfredo. "There aren't any mosquitoes."

After he had slung the fishing line out and tied it to a tree trunk, Wilfredo came back briefly to deliver an armful of dry logs and got a fire going in no time at all. Oliver had the Millbank bag conveniently hanging from a lone tree and soon we had a good hot bowl of spaghetti and cheese sauce as dusk fell. We were to find it was always quite a scramble to get everything sorted out before dark, especially getting into a set of dry clothes while hanging the other lot out to dry round the fire. The problem of drying out our clothes was further complicated by a very heavy dew which fell at dusk and in the early hours of each morning, so that we just had several sets of damp clothes which threatened to go mouldy on us apart from being uncomfortable to put on in the morning.

We sat round the warm fire with the radio blaring so loudly that Nick and I concluded that they must be using it to scare wild animals away. Round about seven Wilfredo and Frank returned with two catfishes weighing six or seven pounds each. Igor was with the two men.

"Wilfredo suddenly just jumped up and ran straight up the beach, pulling in his line," Igor told us excitedly. "Then this huge fish came splashing out of the water. It must have been pretty tough—because Wilfredo had to get his machete to clout it over the head."

We all stood up to watch the moon as it rose, a huge orange circle lifting itself gently up behind the distant line. We found that even for writing we didn't need any other light that night—just the moon and the fire.

Wilfredo was full of enthusiasm for everything, laughing and singing as he joined in helping us amateur campers with not so much as the suspicion of a grumble. He had a running commentary going for Igor telling him all about fishing and hunting exploits, and showing him how to cook the fish. It was just about this time that we discovered that we were almost out of cooking oil and pepper, and very low on coffee. But Wilfredo said he expected us to meet another group of miners higher up and he thought we would be able to buy some from them. This sounded suspiciously like the well-known, ill-founded jungle optimism to me, but I said nothing. Sadly, I was right, we met no one.

That evening we went off to sleep early leaving Wilfredo and Mario preparing fish for a late-night meal, with instructions to wake us when it was ready. We intended to keep up our strength too, somehow.

"But it's delicious," Nick said later, as we ate the meal. He'd somehow managed to have his midnight meal in bed. "Whatever did he put in it to make it taste like this?"

"Well I watched him, and he put in a good handful of pepper, an onion and some comino, and then he squeezed one of those lemons you brought from Lima all over it," I told him.

Oliver was ready for a second helping too, though Katie slept through the whole midnight banquet. Wilfredo had a cooking pot full of fish all to himself and was eating with his fingers attacking the fish's head and tail like we attack a chicken leg, chewing and sucking, quite oblivious to anything around him. Frank was in the same sort of position sitting cross-legged on the beach with a huge plateful in front of him, and a large green leaf beside him on which he was discarding his chewed-up fish bones. The fish had no small bones and we each had several large hunks before finally giving up.

"There's still plenty for breakfast," Wilfredo said. He and the boys all curled up round the fire wrapped in their blankets, while we climbed under our mosquito nets for the remainder of the night.

It was still a dark, dark night when we were dragged from sleep by a loud voice.

"*Levantense!* Time to get up! Come on, wake up."

"But it's only 4.30," moaned Katie. "He doesn't really mean us to get up at this hour does he?"

But Wilfredo obviously did, because all the men were up, the kettle was on the fire and they looked as if they were about to step into the canoe and leave. By the time we had polished off the fish stew for breakfast, and had a lovely cup of hot coffee we were ready to leave, but we had taken an hour, and it was obvious Wilfredo wasn't at all pleased.

Our new Indian Frank was a very quiet person; partly perhaps due to the fact that his Spanish wasn't much better than Oliver's; he didn't have much conversation at all. I repeatedly asked him to tell us if he wanted us to pole upstream any differently, but he just said "*Si, señorita*," and that was that. His face was rather stern and immobile. Only occasionally would he break out into loud, but quite pleasant laughter for something like our frequent tumbles in the river. In a quiet way, though, he was good entertainment value and I think we were secretly very proud to have a real pure-blooded Indian with us, a man from the Mashcos tribe that only thirty years ago had been widely feared for their reputation as head-hunters. In fact the area we were headed for had been too dangerous for white people to enter only twenty years ago, we were told.

The rapids seemed to be coming more often, so that no sooner had we climbed in after pushing up one lot of rapids than we had to jump out and start again. The sun was very hot, and often we walked the beaches while the others punted upstream, so that there was less weight to push, but the beaches were scorching hot and often up to half a mile long. Sometimes we all had to haul the canoes using ropes. Frequently the canoes needed baling out, because of the amount of water we kept bringing in, dripping from our shoes and trousers.

One time when our canoe was up ahead, Frank suddenly called a halt, and we could hear quite clearly the sound of a cat growling, if that sounds possible.

"That's a baby jaguar," he called out. Happily it was nowhere in sight. I don't think he quite knew what he would do with it if he did manage to catch it, and I didn't fancy having one as a pet. Nick was even more interested in keeping a respectable distance from its mother.

About midday we came to the mouth of what Wilfredo identified as the Rio Kipoznue and he called a halt.

"It's time we had a rest. Besides, it's good luck to swim at the mouth of a new river," he said. He dived in, followed shortly by

Wilfredo in Rio Kipoznue
Christening camp — Isaac Margol's place

Lucho. Just as Frank was about to step over the side of the canoe he saw a small stingray. He scared it away with a bamboo pole, then calmly stepped over the other side of the canoe and went off swimming. None of us fancied the thought of swimming among stingrays, and we nonchalantly carried on opening tins of condensed milk which we were having for lunch.

The Kipoznue turned out to be a much smaller river than the Colorado. In some places it was only fifty feet wide, and the jungle seemed to close right in on the edges. Immediately we ran into several sets of steep rapids. We were in front when we heard a shout, followed by a big splash. Looking back we saw Wilfredo appear, clinging on to the side of their canoe.

"That was a spectacular fall." he called out, "That's one bottle of wine I owe but I've lost my cap." Oliver had a handkerchief that would do, and Wilfredo, with his long black, curly hair like a gypsy's, tied this white square, just as old gentlemen do at cricket matches, carefully tying a knot in each corner and pulling it down on top of his head. Katie said afterwards that Wilfredo's fall was quite the best she'd seen, like a back somersault off a high diving board, and he'd been carried downstream some way before being able to stop himself. We now had bets going, for falls in the river. The stake was a bottle of wine per fall and I was well ahead with four—Nick two and Wilfredo one—the others had a clean slate.

The rapids in this section were very tiring and the canoe seemed to be aground as often as it was floating. We were in the water the whole time, never getting a chance to dry out and it was only the sun that kept us warm. In spite of that we had some entertainment that afternoon; Wilfredo called out to Frank:

"Look there's some fish further upstream." Frank leapt out of our canoe clutching his bow and several arrows and made a dash for the edge of the river.

We were told not to go forward, and Frank started stalking the fish Indian style. He moved in a crouch with his arrow strung in

Gold workers at sunset—the last load
13

readiness and crept along the water's edge, sometimes stopping to line up a shot. We saw him loose two arrows, but he didn't spear anything that day. Later on Frank caught fish in this way many times.

At one stage we stopped and collected some turtle eggs, and not long afterwards Wilfredo called a halt, saying that there were no more suitable places to camp so we had better stay here for the night. As we pulled in Wilfredo passed our gun over to Frank and pointed into the trees opposite. Frank nodded and trotted off quietly into the forest. Then there was the sound of a shot and he was back minutes later clutching a wild bird.

The next hour was full of industry with everyone going their separate ways; Katie washing the dishes, Oliver setting up the roofing for our camp, Nick unloading the canoe and me busy making a turtle-egg omelette with some rice and onions, some vegetables, and our last packet of cheese sauce. Mario was rapidly plucking and cleaning the black bird, Frank was off fishing with his bow and arrow, and Igor and Wilfredo were fishing with the line.

We went to sleep very early as we were exhausted and also wanted an early start so that Katie could do some filming. But next morning she was out of luck. It was drizzling, grey and particularly unsuitable for movie-making. And it was especially chilly and unpleasant climbing into our wet clothes. There was no point at all in putting on dry ones when three minutes later we would be up to our thighs in the river pushing the canoes.

Today for the first time we really began to notice that we were moving into higher country. The river was steeper, faster flowing, and there seemed to be a set of rapids at every bend. Above the treetops we sometimes saw misty green hills in the distance.

The stones on the river bed were really big in places. Once I caught my foot between two rocks narrowly escaping a twisted ankle. Mario and Lucho, and most of all Oliver, were beginning to suffer from raw patches on their toes. It was the sand inside their socks rasping and gnawing at their skin as they trudged along the gravelly river bed.

And today was really Stingray Day. Frank had seen one before we even left, just where the boys had been bathing the evening before. He had killed it with a pole and dragged it ashore to chop off the barbed tail. Then he'd carefully picked up the very tip of the tail and tucked it into his bundle in the canoe. Wilfredo told us that the ray's tail was used for preparing a local remedy for toothache.

The stingray itself looked quite horrible. It was roundish and pinky-beige in colour. The tail was about nine inches long, and serrated like a double-edged saw all the way up to the tip. The flat body was about twelve inches across. The tail writhed and struck continuously at the pole that was holding it down like an infuriated snake, while another was bludgeoning the creature to death.

Soon after we had set off Frank saw yet another one and again he jabbed the fish, making it jump so high we thought it might land inside the canoe. Throughout the day there were frequent shouts of "*Raya*", which meant that everyone who was standing in the water scrambled back into the canoe no matter what he was doing.

The stingray has poisonous barbs in its tail. Given the appropriate stimulus—someone treading on it, for example—the ray's tail flips up, doubling back rather like a scorpion's, and inflicting a nasty wound on anything that happens to be in the way. The pain from these wounds, we were told, is quite excruciating. The poison itself isn't fatal, but there's a serious risk of gangrene infection because the barbs are liable to have little pieces of rotting fish caught up in them. Sir Alistair Reid's[*] recommended

* H. Alistair Reid (School of Tropical Medicine, Liverpool), *Toxins of Animal and Plant Origin* edited by A. de Vries and E. Kochva, Volume 3, pp. 957–983 (Gordon and Breach, London, 1973).

remedy for the pain is to immerse the part stung in water as hot as the patient can bear. The pain is relieved within seconds and the part stung must be quickly removed from the water to avoid blistering. It should be re-immersed as pain recurs.

Frank several times stepped into the water without appearing to inspect the waters for *rayas*. When we asked him once if he thought it was safe he just said confidently that there weren't any in that particular stretch of water. How he could be so sure I've no idea, but compared to our ignorance his and Wilfredo's knowledge of the river was mystical, almost supernatural. They always saw *rayas* long before us no matter how hard we tried to outdo them. They would scan the surface of the water reading it like a book, calling out if they saw a shoal of fish twenty or thirty yards away.

"Quick, Frank, bring the gun," called back Wilfredo once. They both dashed off splashing through the knee-deep water up ahead, then wading in to waist level, holding their guns high while they pushed their way into the centre. At this particular section of the river the jungle came right to both banks, hanging over the edges, and it looked a particularly murky, snaky, type of water to have to wade through.

"What are they trying to get?" Katie asked.

Oliver called out, "I saw something swimming in the water but it went into the bank."

"It sounds like an otter to me," said Nick, rightly as it turned out. "Let's hope they don't get it. They only want the skin; it probably has some value in Puerto Maldonado otherwise they wouldn't waste all this time."

Happily they never did get the animal.

We came to two very difficult patches. The river had split into several very small creeks and we all had to pile out while the boys cut a passage through fallen branches with their machetes. Just at this narrow passage the river water gushed through waist high, creating a sort of low waterfall. Igor climbed to the top and sat down to come wooshing gaily to the bottom, using it as a slide, while we all had a welcome cooling off dip. But pushing the laden canoes up this section wasn't so much fun. Soon afterwards we came to yet another obstacle. It looked impassable to me, but Wilfredo ordered everyone out of the canoes and set off with his hands clasped behind his back like a general about to inspect his army. There were two very large tree trunks lying parallel, close together, right across the narrow river and just under the surface. There was no way round and no way through, and no chance of floating the canoes across. The blockage caused by the trees made the water come surging through beneath so fast that it was difficult to stand up.

Wilfredo scratched his head, muttered something about wishing he'd brought some dynamite and walked round the obstacle, climbing on top of the tree trunks and then came back to the canoes.

"Okay. Now everyone help first with this canoe. I'm going to push from behind, we'll try and get the canoe half across then we'll all put our weight on the other end and see if it will lift up slightly then we'll PUSH."

Katie and I could barely keep upright in the current, so we just watched the whole operation while the boys got to their respective positions, lining up the canoe. When Wilfredo shouted they charged forward, three hauling with a rope and two pushing from behind. The boys up at the front struggled in water up to their armpits. It took them something like an hour to get both canoes past this obstruction so when they were finished it was time for some cold rice and tuna fish. After an all out struggle like that the men were really ravenous, and wolfed everything down until there wasn't a grain of rice left.

The Rio Kipoznue was a very strange river; its character changing continually as we progressed. In some places, like the spot where they had been hunting the otter, the river was still, deep, muddy and overhung by the jungle. At other times it was all stony, open beaches and big sandy patches, with the river very shallow and fast running.

Often during the day Wilfredo called a halt while he and Frank went off to fish with the bow and arrow. They were after a species called *boca chica* or small mouth. This was a fish that lived through sucking in algae from the river water; it couldn't be caught with a hook and line as it didn't eat the bait. One way to catch them was to throw a net, but Wilfredo's didn't have enough weights on it, so it was left to Frank and his bow and arrow. We would get rather impatient because we hated eating *boca chicas*, they were full of tiny bones. Here we were, having to get up before dawn each morning so that we could cover a good distance during the day, and now we kept stopping for hunting and fishing sprees. But at times we were grateful for these delays; it was pleasant to have a rest.

After another long day in the water Olly's toes were quite raw at the tips and were becoming a real worry. He promised faithfully that he had put on foot powder every day and the only other thing I could think of was to coat them in antiseptic cream. Lucho and Mario had the same thing to a lesser degree so powder and cream did the rounds that evening. Nick had a very nasty sore on his finger from a burn. As all these maladies were being soaked in the river all day none of them had time to dry out and heal.

Just as a steaming hot cauldron of soup was ready, that evening, the heavens opened. We all made a dash for shelter under our plastic sheeting but still everyone was soaked. It cleared up enough for us to have a second meal of fish at about 10.0 p.m. but next morning it was drizzling steadily. We took one look at the weather and decided to sleep in and give it a chance to clear up.

But it didn't. Today we had to pull on our wet clothes without even the sun to cheer us up, and soon after we set out the drizzle gave way to a steady downpour, broken by intermittent dry periods. During one of these clear times we saw a deer by the riverside. We'd been eating nothing but fish for many days, and everyone was yearning for something different, but Frank's rather hasty shot missed altogether. We groaned and wore long faces for a while after that.

At one point Wilfredo went off with Lucho, Frank and Mario to inspect the lie of the land. We stood waiting for them in the middle of the shallow river using our wooden pans as hats to try and ward off the rain, though we were so wet by then that it was rather pointless trying to protect ourselves.

"Even I'm shivering now. Don't you think we could make a stop here?" asked Oliver.

But there was nowhere we could set up camp without the risk of being washed away if the river came up. Oliver announced that he was going off for a walk on the beach to warm up. The rest of us raided the green canvas food bag and ate peaches and condensed milk, passing the tins round and savouring each mouthful as though it were our last, as indeed, peaches-wise, it nearly was. All at once the absurdity and discomfort of our situation was too much for us. We all looked at each other, with the rain pouring down solidly, and began to giggle. We spluttered and shrieked with laughter, choking on our precious condensed milk, for several minutes. Drowned rats never looked as wet or as ridiculous as we did, standing knee-deep in a small river two hundred and fifty miles from the nearest town, with rain dropping from the sky like a waterfall, wearing wooden pans like Chinese coolie hats upon our heads.

When Olly came back he announced to the world in general that these were the very worst few days of his life, bar absolutely none. We had known he was pretty miserable because Lucho was being so unhelpful, but Oliver was so quiet and stoical normally. Things had to be really bad for him to speak so vehemently.

"If there was a helicopter round one bend and Fort Knox with the doors open round the other, I'd just take the helicopter," he said.

"But, Oliver, surely when you sailed you must have been frozen and soaked to the skin hundreds of times," Nick said. He was himself about to set out to take up sailing seriously and he thought he could guess what it was like.

"Of course I've been soaked before," Oliver said scornfully. "But at least I always knew then that I could get home to a hot shower and warm bed within the foreseeable future."

He had a point there—we were a long way from such luxuries. There was a general silence as we all fell to dreaming of home comforts. To cheer ourselves up we raided our food bag yet again, this time committing a scandalous indulgence: a whole bar of chocolate each, something quite unheard of. Normally each bar was rationed out painstakingly, square by square. All Nick could say rather lamely was:

"Somehow we must all have been in as bad a position as this before now. This time at least we do have food!"

Soon the others returned and we set off again but about mid-day, as soon as Wilfredo saw a suitable place he called a very welcome halt. Even he agreed that the weather was far too nasty for us to continue. We weren't just wet, we were wringing wet, and as soon as the boys had constructed a shelter of bamboo leaves and Wilfredo had a fire going—getting a fire going in a thunderstorm was something else he could do—everyone wrung out their shirts and stood over the fire, eyes streaming from the damp wood smoke.

"I've seen more shirts burn that way . . ." said Nick, as Lucho held his T-shirt right on top of the flame to dry it out a little. But Lucho was careful or perhaps lucky, and there was no mishap. In fact Nick himself ended up the trip with all his clothes singed or burnt. We ate scrambled turtle eggs with some vegetables mixed in, spooning them straight out of the frying pan. Afterwards we had hot coffee, and by then all of us were starting to feel a little more human again.

The weather cleared up in the afternoon, and the sun came out, but Wilfredo decided we should all get properly warm, dry our things out and hope for an early start next day. Even our bedding felt damp, so we laid everything out across the beach, covering practically the whole area. Then Katie and I sorted out our cameras and film and the boys relaxed with a game of cards.

After dusk we were sitting near the fire having our evening meal with our plates resting on our laps, when suddenly Oliver and Igor beside him leapt to their feet simultaneously shouting:

"Run. Watch out! A snake!"

Everyone scattered. I saw nothing, and heard only a rustling sound as it slithered off behind our makeshift kitchen. Wilfredo was at the spot within seconds, holding his shotgun, but it had already vanished.

What colour was it? How big was it? Who saw it anyway? The questions poured out as we milled excitedly around the calm figure of Wilfredo.

"I saw it," said Igor. Mario and Frank agreed that they had seen it too.

"It was coming straight towards me," said Olly. "Then I jumped up and it veered off. I think it was as surprised as I was."

"It must have been six feet long," said Igor. Mario and Frank agreed that it must have been. Frank added that it was a dark colour with a lighter underside.

"Who saw it first?" asked Wilfredo and Olly spoke up. Wilfredo shook his head sadly. "That's a bad omen. My sister died four days after a snake crossed my path. Yes, it's bad all right." Oliver hadn't quite grasped the import of these doom-laden Spanish phrases, so we hastily changed the subject.

One thing we'd discovered by now — Wilfredo was very superstitious. The previous day Mario had seen a turtle swimming downstream. He and Frank had dived for it and—rather impressively—managed to catch it with their bare hands. We'd proudly poled up to the other canoe saying, "Look what we've got for supper."

But Wilfredo had been very angry.

"Never catch a turtle while we are on a journey—it's sure to bring bad weather. Put it back right away." To our chagrin Wilfredo was quite adamant on the point and rather than cause a scene we had to abandon one of our favourite meals and watch it swim happily away.

We all went uneasily off to our sleeping bags wondering if that snake was still around. Wilfredo assured us that nothing would attack a mosquito net, not even a jaguar, but it didn't stop us all tucking our nets under with special care that night. It was good that Oliver and Nick had the two outside places, I thought!

Next morning was dry though not sunny, but anything was better than that torrential rain. We poled for several hours and then Wilfredo called a halt. He jumped out of the canoe, picked up his shovel and pan and told us to follow him.

The canoes were wedged on a shallow piece of rapid so we abandoned them in midstream searching under the plastic sheeting for our pans. We dug where Wilfredo had started and got on with our panning, watching to see Wilfredo's technique and busily copying him. Wilfredo had dug near some clay which he said was where the gold ought to be.

It was Mario who got the first piece. We all dashed to look at his pan, and there right in the bottom was a piece of gold the size of the "o" in this type-face.

"Oh, but they can come much bigger than that," said Wilfredo, dismissing our excitement. It seemed quite big to us, as we'd only seen tiny powdered specks, down in the Colorado River. "Anyway," he went on, "this isn't the place we've been heading for. It's much higher up, come on let's go." Olly had found a good big speck himself, so he was loathe to leave.

"This is the kind of panning I've always wanted to do," he said, looking a lot happier than he had the day before.

So we pushed on once more, hopping out often to struggle through what were now much stronger, steeper rapids. Wilfredo assured us that six bends more and we should hit a creek nicknamed Rocas—literally rocks, which was one day from the place we were planning to make our base camp. But the day wore on, and there was no sign of Rocas. Our tempers started to fray.

"He's been saying six bends for the last twenty bends. Can't we at least stop for something to eat," said Oliver. But Wilfredo pushed on and the second canoe, with Nick, Mario, Frank and I, fell back quite some way. At one narrow set of rapids we nearly had a bad disaster. The canoe somehow got sideways and snagged on a rock. It started filling with water. None of us was prepared for this, and though we tugged desperately trying to haul it up, it wasn't any good. Frank saved the day by grabbing the rope from Mario, pulling the prow round and letting the canoe float gently back down the rapid, so that it faced the right way again. After this we were really exhausted and very fed up. After baling out we pressed on to find the others waiting for us enjoying a snack, this time of tuna fish followed by tinned pears which made us feel quite different people again.

At a place where the river was deep and very still Wilfredo pulled in to the side. This was an abandoned field, from ten years back, he told us. He called to Frank and Mario to join him with their machetes.

What a resourceful fellow we had with us. Ten minutes later they all came back laden with green bananas. The plants were now semi-wild, and the bananas were much smaller than the normal cultivated ones we'd eaten before. But what a treat!

We'd all been feeling a bit peeved at Wilfredo for making us think we had nearly finished our journey when we were still miles away, but now our annoyance evaporated, and we were all smiles.

And a few minutes later we were positively excited, when Wilfredo jumped out of his canoe and announced: "This place is called Bandera. We've arrived."

CHAPTER SEVENTEEN

Bandera

Goodbye

NICK WAS STILL SCEPTICAL.

"You mean we've arrived for tonight, or we've actually arrived?" he said. "I thought we had to get to Rocas first."

"We're here—this is it," explained Wilfredo. "We must have passed Rocas without my noticing it. It's been years since I came up here, and these rivers change so much."

This was the place where years ago fifteen canoeloads of people had come to prospect for gold. The pickings had been good enough to keep the same people coming back each year. The Spaniards, too, had come to this river for gold, and before them the Incas. But the physical difficulty of working in this remote spot had eventually overcome the miners, and these days nobody came.

"There were too many hardships," said Wilfredo. "We could never get enough supplies up here to feed everybody, so in the end we stopped coming."

Bandera. The name means simply flag in English. It was called that because the first man to arrive there one year had stuck out a flag at the entrance to the little *quebrada* where we were to work, so that all the following canoeloads of people would know where to go.

Wilfredo was very firm that we should find the highest possible position for our camp. "This river rises in just a very few minutes. If that happened we could be in real trouble."

So we found a spot high up on the beach and tucked right up against the jungle. It took us about an hour to clear the area, and to build bamboo frames for the shelters. Mario got a fire going. We hung up our damp bedding and mildewed mosquito nets, and shook out our bundles of muddy, smelly clothes. It was wonderful to be settled again with a prospect of relative comfort, knowing that we didn't have to turn out at dawn for another hard day's slogging up the river. Now we could look forward to getting at the yellow stuff.

Our new camp was located right on a bend, so we couldn't see much of the river upstream. The view downriver was something special though. The river stretched out below us, punctuated by three sets of rapids winding gently before it curved out of sight. Beyond lay a panorama of trees mistily silhouetting the sky in delicate hazy pastel shades in the distance. What an exhilarating change to be in hilly country after so long in flat, featureless jungle. With our noses in the water while we had pushed and hauled the canoes upstream we'd hardly noticed how high we'd come. We were now about eighty miles from the nearest group of people we'd passed previously. It had taken us six days to get here instead of the eight originally predicted by Wilfredo. Probably it had been so long since he had been here that he'd just forgotten exactly how long it was likely to take. Anyway no one was complaining—I think two more days' travelling would have finished us all off.

In the morning Wilfredo sent Frank off hunting. Igor and I stayed at the camp to have a rest, while everyone else set off by canoe, eager to search for gold. Frank came back at midday with the news that he hadn't killed anything yet. After a short rest he set off again in a different direction. I remember thinking he looked very odd indeed, armed to the teeth with a shotgun plus bows and arrows, ragged and barefoot yet crowned with a neat white hat like a cricket cap. We became rather fond of Frank, he was our silent watchdog. He would sit by the fire for hours on end, purring like a cat, looking as though he were in conference with the spirits. Maybe he was.

Oliver came back soon after Frank left. He wasn't having any luck, he said, and he had a sore finger which he thought might clear up more quickly if he let it dry out for an afternoon. We all had little things wrong with us: I had a swollen hand, for no apparent reason; Katie, Oliver and myself had all developed puffy eyes, especially in the mornings. Wilfredo attributed this to the continuous smoke from the fire, but we weren't so sure. We'd now been cooking with a wood fire for more than a month, and though we'd often had streaming eyes from the smoke we'd only recently begun to look as if we hadn't slept for weeks.

Everyone came back from the first day's panning towards dusk. The day had been a partial success, though more tiring than fruitful. They had all set off after Wilfredo and Mario following Bandera creek, sometimes climbing round the edges into the jungle when the pools were too deep, but mainly walking along the creek bed itself. They were only able to keep up with Wilfredo and Mario because every now and then the pair would stop to have a dig and see if they had a good patch. They had found one or two chips, but nothing really exciting. Still, the gold fever had evidently set in and Wilfredo said they had a good spot for us to start working tomorrow.

Frank arrived soon after the others bearing three blue and yellow parrots. Two were ready for the pot, and one was alive with an injured wing. Frank wanted it as a pet, so it was tied up in a tree by the camp, where it lived for the next few days, stripping the bark, cackling and generally adding to the atmosphere of the place. It eventually went back to the village as Frank said he was breeding them. He had also caught several fish with his bow and arrow. But Wilfredo wasn't at all pleased.

"Look at all the cartridges he's used today, and there's hardly enough meat to go round. He should have gone for bigger game." Mario and Frank began deftly plucking the dead birds, and soon the camp was littered with brightly coloured feathers. We would be having fish for supper though, because as Wilfredo explained, these birds could easily be twenty-five or thirty years old and were likely to be as tough as old boots. They needed to boil all night on a low fire. Frank dangled the scrawny birds over the fire to burn off the remaining feathers then he cut them into pieces, putting some of the entrails to one side. Later he wrapped these in a leaf and baked them in the fire. The rest — including the head and feet — all went into the stew, with a few handfuls of dried onions, comino, pepper and salt, plus plenty of chopped garlic. The fire was stoked up well and Mario was elected to tend it, refuelling it two or three times during the night.

Wilfredo was continuing to keep us entertained with a stream of jokes, singing and general energy. He never seemed to tire, and his blaring radio and endless late-night card games with Lucho continued to be a feature of life after we'd established our new camp. When he wasn't busy with something else he would sit down and sew his clothes or go off to the river to wash them, though Katie had repeatedly offered to do his sewing for him.

"I'll do it myself," he had insisted. And when Katie offered him some tough nylon thread instead of cotton, he shook his head.

"What's the use of having something that won't break? I wouldn't have any work to do!" And he carried on singing to himself as he sewed away, sitting cross-legged like an old-time tailor on the beach.

In the morning as there was fish for breakfast as well as parrot I plumped for fish. I had a quick taste of parrot just to say I had eaten some, but it tasted rather powerful for so early in the morning. Katie, not being a fish eater, had two platefuls of parrot, and Frank ate the heads. These were obviously a great delicacy because he offered them politely to everyone before putting them on his plate.

We set off for the day, poling upstream a little way before turning off the Kipoznue into little Bandera creek. There we left the canoe and set off on foot, following the most beautiful stream I have ever seen. Long overhanging boughs stooped to touch the sparkling water as it rushed rapidly and noisily over the stones. In places there were deep mysterious pools and for much of its length the creek was in a narrow canyon where only slender shafts of sunlight could filter through the dense forest and the air was cool and pleasant.

Wilfredo and Mario had gone ahead. When we finally caught up with them they had handkerchiefs laid out on the stones. Already they had accumulated tiny piles of gold chips in the centre of their cotton squares. We'd discovered that the fashion up here was to collect our gold chips in our handkerchiefs during the day, and when work was finished we pushed all the chips to one corner, knotted the handkerchief round the chips and then tied it round our neck. We ourselves now settled down to try our luck. We were working on a dried-up branch of the stream, which was littered with large, heavy stones the size of footballs. We cleared this surface away, looking for the clay that Wilfredo told us was the most likely place to find gold. The clay itself was hard, and trying to push the shovel into the stuff was exhausting work. It took four or five shovelfuls to fill our pans with a mixture of fist-sized stones—we would pick out anything larger by hand—gravel and pebbles of all sizes, sand, a little black sand and perhaps a chip or two of gold. Every now and then Wilfredo —between singing and directing Mario and panning himself— would come over to our patch, and advise us where to dig, sometimes putting a few shovelfuls in our pans, which was like giving us a lump of gold itself.

The serious business of panning turned out to be backbreaking work. By now I was quite tough and fit, but I found I just couldn't squat holding this very heavy load panning away for more than ten or fifteen minutes. Like everyone else I had to take frequent breaks, and I calculated that Wilfredo was washing four pans for every one that each of us did.

With all that effort it was quite exciting when someone had a chip. We soon gave up dashing over to see Wilfredo's or Mario's chips because they appeared rather too frequently, but when Katie or Nick, for instance, got near to the end of a pan we would all crowd round waiting for the very last bit. There was quite a bit of suspense involved, because we had to wait until the final grains of black sand had gone before we could see what our hard work had produced—if anything. When a gold chip appeared it was such a boost that we just had to get on quickly to see what our own pan would produce. But the work did need some dogged persistence because I worked a whole morning without getting anything in my pan. Igor too had to stick at it for quite some time before he came up triumphantly with two small chips in the same pan. Meanwhile Wilfredo and Mario had several times found three or four chips in one go. These chips were between a ladybird and half a thumbnail in size and they were flat, battered by crushing between rock strata in the motherlode, and pounding by fast running waters on their way downstream.

We worked hard all day from 7 a.m. and arrived ravenous back at the camp at around five, when it was nearly dark. That night when the rest of us were asleep Nick got up to walk down to the river's edge. He was startled when Frank loomed from nowhere out of the darkness with a shotgun in his hands, whispering, "*Sachavaca, señor*. There is a tapir around." It seemed that the tapir had lumbered straight through the camp as though we hadn't existed. I can imagine the panic and consternation if the animal had come blundering into our shelter. If there'd been a moon that night Frank would probably have been able to shoot it; as it was the tapir had a lucky escape.

We panned on conscientiously for a few more days. Then Nick

announced that he could postpone the fateful day no longer; his backlog of work would be piling up in Lima, and Igor had to go to school; they both had to leave. Meanwhile, Katie and Oliver had plane reservations to Miami in two and a half weeks' time. Katie had to get back for her wedding in Switzerland and Oliver had had enough. It was time for them to leave also. Me, I just couldn't pull myself away from hunting for gold a while longer. For me the whole business was as compulsive as sitting at a roulette table. Sooner or later I *knew* I was going to find a nugget the size of my fist. And, as if I needed any extra incentive, Wilfredo said that further upstream there were some petroglyphs and ancient ruins, which I wanted to see more than anything. I elected to stay on. It was decided that Frank and Lucho would get the others safely back in Frank's canoe, and Wilfredo, Mario and I would stay on looking for gold and Inca ruins.

So the day was set for their departure, on August 23rd, nearly eight weeks after we'd first set off from Puerto Maldonado. Before they were all due to leave we had a special feast of wild turkey and rice. Thank goodness Frank had managed to shoot something more appetising than parrot for this meal. There was a bottle of wine, too, which we had saved for a special occasion. This seemed to be the right moment to drink it. We rushed round taking last-minute photographs and filmed a bit and I wrote a couple of notes home, to be posted from Lima. Everybody by this time had a few chips to take home with them but they weren't exactly leaving with their fortunes on their backs.

Then they all piled into Frank's canoe and poled off, waving like mad. Their departure seemed almost abrupt. I hadn't been thinking about it at all, and suddenly here they were all leaving. I watched the canoe as it grew smaller and smaller, floating away down that long stretch of river below the camp. Then it disappeared round the bend. I felt very sad. Katie and Olly and I had been together for nine weeks. Inevitably, we'd had our differences, but they had been small and had never got out of hand, so we had always been able to laugh at ourselves after an argument. Most of the time we had got on together really well.

Oliver had had the most to put up with. Coping with two girls sometimes put Olly out on a limb, and it might have been better if we'd had another man along to give him moral support. He was the one who worked the hardest, both on the golding operation and seeing that the camp ran smoothly — collecting wood for the fire, seeing water was always filtering and generally organising the various camps we set up. It was only in rare moments of stress that he would give vent to his feelings of frustration and discomfort. Deep down Oliver was a clean air, cold climate, mountain and sea man — the jungle hadn't lived up to his expectations and there was nothing to be done about it.

Katie was always helping to get something done. She'd be quietly and efficiently seeing to things that needed doing before we had even noticed, so that we were always in a tidy state which certainly wouldn't have been the case if it had been left to Oliver and me. And she was so kind that if things were going wrong just having Katie around meant a lot to the group. Thanks to her we were stoked up with pills unfailingly each morning, and our vast assortment of film survived in perfect condition because of her ministrations. She had to get through many bad patches, particularly towards the end when almost our entire diet for days would be fish. She could not stand even the smell of fish, so she had to plod on with a very tedious and unsustaining diet. Katie had wanted to stay on with me but sadly she couldn't. She had a definite plane to catch but Wilfredo was being rather vague as to how long he wanted to stay on the Kipoznue.

"It depends on the gold," was all he would say. And his canoe was the only available transport back to civilisation.

CHAPTER EIGHTEEN

Kipoznue

Gold and Ocelots
A Christening

MOST OF THE DAY had gone by the time the others had packed up and left, so there was no time to walk up to our usual place and do some panning. Wilfredo decided to go off and try out some patches for gold closer to the camp, and I got on with a much-needed hair and clothes washing session.

That evening seemed very quiet, a bit flat, almost forlorn with only the three of us in the camp, instead of the usual boisterous crowd. I chatted with Wilfredo about this and that, and he asked me what England was like. He had once seen an old film set in London, one of those productions which has the city permanently wreathed in smog. He wondered how we could bear to live in that climate. I told him it wasn't like that all the time and he was particularly impressed by the thought that parrots cost £60 each in Harrods. He didn't much like the sound of our very short winter days, either.

I stayed up late writing with only the crickets squeaking and the tree frogs croaking to keep me company. Mario had fallen asleep curled up by the fire, as he'd been told to keep an eye on the dried beans which had to boil all night to be edible for the morning's breakfast. Wilfredo had stood up early on and announced that he was off to have a snooze. Once a nearby tree came crashing down, and we all dashed out to see what had happened, shining our torches. But no more sound was heard, and we withdrew to the camp.

Sleeping on the ground was still, for me, desperately uncomfortable. I'd have a sore back through panning all day, so I would just lie there shifting around, looking for a nice position and never finding it. I wasn't getting enough sleep, though the actual hours of rest—practically dusk till dawn—were enough for me never to have to rise feeling exhausted. In the morning there was compensation; I had just to lift the corner of my mosquito net at dawn to have what I soon began to think of as "my view" downstream, and there couldn't have been anything much more beautiful than waking up to the mist over the trees and the early light sparkling on the rapids. The early morning smell too, fresh and a little damp, added to the magic so that I couldn't quite pull myself out of my warm sleeping bag until the spell was broken, when one of the boys switched the radio on or called out that breakfast was ready.

I was a little nervous going off panning the first day I was alone with Wilfredo and Mario, because they habitually sped along so fast, as they made their way up the creek, that I had great difficulty keeping up. I seemed forever to be stumbling and slipping on the mud and the wet stones. But it wasn't always me that slipped. Once while Mario was climbing down a steep bank, he tripped over and dropped his machete. It slithered down and fell into a deep pool. Wilfredo stopped.

"Better wait for him in case there's a jaguar around," he said with his usual good humour. Wilfredo had the only shotgun now, and he was never without it. He even slept with the gun on one side and a handy pile of cartridges on the other.

When we started panning I felt a bit helpless about where to dig. I still didn't know enough about gold mining to pick out the best spots. Most of the stones were too heavy for me to lift on my own, as well, but Wilfredo soon put me at my ease by showing me exactly where to dig. If my luck was poor he would tell me to go and dig in Mario's or his patch, whichever was doing better. In the end I was getting quite a few chips, not nearly as many as Mario, who could work four panfuls while I was still on my first one, but at one point I started outproducing Wilfredo. He went through a bad patch, and after a while his cheerful singing tailed off. At length he gave up, saying he was off hunting. Later on, though, we heard him working a little further downstream, whistling away, and he reappeared to accompany us home before it got too late. Walking back I was saved from my regular worry of keeping up because they both tried their luck fishing in each of the deep pools we passed. Eventually we had about a dozen little five-inch fish for our supper and for bait. Wilfredo announced that we would take the next day off, because he wanted to go upstream to see if there were any ocelot tracks. I decided to go with him, and Mario would come too.

At dawn we heard a sound coming from the forest nearby, rather like a person snoring. Mario sat up.

"Turkey," he said. Wilfredo, though, had made it quite plain, several times, that fish was his favourite dish and in his opinion birds didn't have nearly as much goodness in them. He lay in his mosquito net for ages before suddenly getting up.

"I suppose I'd better get it," he sighed, in an annoyed sort of way. "It's so near." Three minutes later there was a bang. He came back clutching the turkey, and slung it at Mario, telling him to clean it.

We set off poling upstream in the canoe, while Wilfredo walked along the beaches searching the sand for animal tracks, jumping back into the canoe when the forest came to the river's edge. He called out several times that the area was full of *tigrillos*. Mario and I poled lazily up the river, stopping to fish at the deep pools. We came to a place Wilfredo said was called Torre — meaning tower. It was a beautiful canyon with the rock face rising steeply straight up for several hundred feet. A few trees clung precariously to patches of hard clay as if just a little breeze would force them to let go. On the opposite side was a small steep beach from where Mario and I fished.

"There's so many *tigrillo* tracks here we could stay a fortnight just building traps and I could make more money than with gold," Wilfredo announced. He went off fishing with his bow and arrow leaving instructions for Mario and I to fish in a particularly deep pool. The sun was boiling hot and though I started off full of good intentions about doing some serious fishing, after a while I sat down in the sand. Then I must have lain back, because I dozed off, still clutching my line. Suddenly I heard Wilfredo shouting excitedly from the opposite bank.

"*Jala! Jala!* — Pull!" Mario was running as fast as he could up to the top of the beach though he was obviously having trouble doing so; something was holding him back. I dashed across and we heaved together running up the beach, pulling with all our strength till an enormous fish came flopping out of the water and lay gasping on the beach. It was a catfish, quite the largest we had caught to date. It must have weighed thirty pounds and was about three and a half feet long. I put my fishing line away, thinking that the fish would last us a week so it was pointless trying for any more. But not at all, said Mario, we could salt and dry out the fish which would keep for months, and so we might as well carry on fishing.

A few minutes later it was my turn to hook one. I ran up the beach battling with a very active fish on the end of the line. This one was a lot smaller, weighing about ten pounds, but I was very proud as it was the biggest fish I'd ever caught by myself in my whole life. It was a *sábalo*, said Mario, one of the best ones to eat fresh. We would dry out the catfish. Soon Wilfredo himself came back waving the bow and arrows which Frank had left behind for us to use, and he was clutching a large fish. He was grinning from ear to ear.

"It's a long time since I've been fishing with a bow and arrow but I don't seem to have lost my touch."

His fish weighed around seven pounds. He had also shot another wild turkey. I just wished the others had still been here to help eat everything up. Often since they had left us I had wondered how they had got on with their journey downstream. Mario kept a sort of running commentary with, "Well, they must have passed the mission by now," and "They should be at the mouth of the Colorado." There was no chance of hearing news of them over the radio, because we couldn't get a peep out of Radio Madre de Dios here at this distance from Puerto Maldonado.

Wilfredo just said, "We'll know soon enough when we go back down the river."

We paddled back to camp, passing the sandy beaches Wilfredo had walked up that morning.

"Always go silently past these beaches in a new territory," said Wilfredo. "On a good hot day like this jaguars like to sunbathe in the sand." But we never did stumble across one.

Back at the camp we sat down to devour a pot of wild turkey that we had prepared before leaving in the early morning. Then we started the really hard part of the day's work, cleaning, preparing, salting, curing and cooking our haul of fish and fowl. Now I discovered why Wilfredo had carted a whole sack of lumpy salt up to the camp with us. He put the salt down beside me and told me to grind the coarse crystals down to a fine powder between two smooth stones from the river. In the meantime Wilfredo and Mario got down to the dirtier business of plucking, gutting, scraping and slicing. We kept my *sábalo* for the pot and stacked the rest of the fish in a pile of thin slices with liberal layers of crushed salt in between. The hot sun of late afternoon began the process of drying out and curing the fish. One more day of sun would finish the job, Wilfredo said. But all our efforts were for nothing. Next day was cool and overcast, not hot enough to dry the fish. The day after Wilfredo and Mario tried the meat, and with many "Ughs!" and cries of disgust, pronounced it *malogrado*—rotten.

Mario was a small boy who looked about fourteen years old, though he was actually seventeen. In spite of this he was strong and worked as hard as anyone I've ever seen, under the watchful eye of Wilfredo. Wilfredo was an active and hard-working person, and he was determined to have his workers in the same category—whether they liked it or not. Each morning he would turn Mario out of bed before dawn to get the fire going and put the breakfast on. Sometimes Mario would be so tired in the early evening that he would fall heavily asleep by the fire. The only way Wilfredo could wake him was by shouting.:

"*Creciente! Creciente! La canoa—corre!*"—The river's rising! Get the canoe—run! Little Mario would leap up automatically and dash off half asleep to check that the canoe was securely moored. This happened several times when Wilfredo couldn't wake him by any other means.

Ever since we'd been on the move from the Rio Colorado I had taken to sleeping fully dressed inside my sleeping bag. Nights were a little cooler at this altitude—about eight hundred metres above sea level—and in any case, Wilfredo had drummed it into us many times how dangerous it could be if we were caught unprepared with the river rising rapidly.

"It could cover our camp in fifteen minutes," he had once said.

And the precaution was to come in useful this next morning. So early that it was still almost pitch-dark Mario came running and announced in a loud and breathless stage-whisper that there was a tapir near the camp. Instantly Wilfredo snatched his shotgun and vanished into the darkness. Seconds later there was a bang. He came back signalling urgently to me.

"It's wounded. Are you coming?" I was already out of bed. I picked up my boots, dashed down the beach in my socks and jumped into the canoe as they pushed off. We could see the huge animal clambering out on the far bank. As we hit the shore the other two scrambled off into the undergrowth while I sat there blearily trying to tie my boots on.

Soon they came back.

"It's gone downstream. Come on, hurry, Mario, push off," shouted Wilfredo. We floated down, and Wilfredo stood up in the bows, his sharp eyes scanning the bank for signs of the wounded animal.

"There it is! Pull in, Mario, hurry. *Burro* — donkey, you're so slow!" Wilfredo was so impatient this morning. He jumped out. I didn't follow him because I didn't want to spoil his hunting, but he waved irritably, signalling for me to come on and be quick about it. I raced up the beach after Wilfredo. The tapir had swum downstream and was now making off across the open beach, heading for the jungle. Wilfredo fired twice, hitting the animal both times. But it was a huge thick-skinned creature, difficult to kill, and Wilfredo's cartridges were only loaded with light bird-shot. The wounded beast doubled back and came lumbering up the beach straight towards me, then Wilfredo caught up with it, and put a final shot into it at point-blank range. The tapir fell and died, almost at my feet.

It took us fifteen minutes to drag the carcass to the canoe, a distance of about thirty yards. To get it aboard we had to tip the canoe right over on its side and shove the tapir over the gunwales. I think it must have weighed at least four hundred pounds. It was basically a rather repulsive-looking creature, with tiny eyes, a fat and piglike body, browny and with few hairs, short spindly legs and three-pronged hooves. Its redeeming feature was a long, lugubrious face, which made it look somewhat pathetic and appealing.

At the camp Wilfredo turned the canoe upside down in the water, and they used the flat bottom as a chopping block, cutting the great carcass up into small pieces for salting and curing. Wilfred said the dried meat would keep for up to a year, and would be really useful for feeding his men. We put in a whole day, carving up the animal, grinding salt and preparing it.

"What we need now is some sun," said Wilfredo anxiously. He decided to prepare a store for the meat that would let it keep for two days. We hoped we would have sun by then. On the ground we laid out enormous leaves taken from a tree near the camp, then wrapped all the meat on a blanket of leaves. A final layer of leaves went over the top of the pile, and we weighted the whole lot down with stones.

"The important thing is not to let the air in," said Wilfredo.

That night we ate fresh tapir meat, small chunks deep fried in tapir fat. *Chicharrones*, a Peruvian dish. They came out hot, almost crunchy, and utterly delicious. A nice bonus from this meal was a big pot of lard which we could use for frying. We'd run out of cooking oil two weeks earlier, and had been eating stews solidly ever since.

The weather still looked gloomy that evening. Wilfredo predicted rain and said he would go off with Mario to build ocelot traps next day rather than squat shivering in the river. This was a disappointment because I wanted to do more panning. We had left our pans up at Bandera creek and I was much too nervous of the jungle to go and fetch mine on my own.

Next day it did rain—all day. The two boys went off to build traps, and I did some fishing, without any luck. Wilfredo had shot two monkeys that morning for baiting his traps. For two days they built traps, completing four in all. I went off to inspect them when they had finished. I had made it plain that I didn't approve of hunting these animals for their skins, but I didn't labour the point. This was Wilfredo's way of life and had been for years: lectures about conservation and endangered species were incomprehensible to him. I think the real fault lies with the women who buy coats made from these skins, and the countries that allow the coats to be sold in their shops. Each skin would net £30 — to Wilfredo — the equivalent of three or four hard days' work for gold. If there was no market for the pelts Wilfredo and people like him wouldn't bother to shoot and trap them.

The traps were very cleverly made. They were a kind of cage, very solid, made from roughly hewn logs laid one on top of the other and supported by vertical stakes set into the ground at the sides. A heavy log hanging vertically over the entrance served as a trapdoor. The log was strung up over a crossbar arrangement made of bamboo poles and held up by a rope running down over the top of the trap to the back, where it was passed through inside and wrapped round the bait, a dead monkey. If a *tigrillo* tried to grab the bait this rope would snap loose under the weight of the trapdoor, which would come down with a thud behind the animal. The design included a floor of branches to stop it from burrowing its way out.

They had had some difficulty in finding the right kind of root to make the rope for lashing everything together. Wilfredo said that the only root strong enough was that of the *tamiche*, a parasitic plant that grows high in the trees of the forest. The plant sends down long, long filaments which eventually reach the ground and take root there. The rope was only strong enough to use if the plant had reached the ground and rooted itself.

"It's the plant that Tarzan swings on," said Wilfredo. He was a big fan of Tarzan.

Our supplies were beginning to get rather low. Sugar was especially short, and we were now permanently drinking tea, as the coffee was finished. On top of this our efforts to salt down the tapir meat were a failure. The weather was against us, and there was no sun to dry the flesh out. I spent a whole day trying to smoke the meat, but in the evening Wilfredo's verdict was that we hadn't used enough salt to preserve the meat. One sunny day and everything would have been fine, but as it was the meat had gone off, and there was nothing to be done but throw it in the river for an alligator we had noticed frequenting the river a bit further downstream. I was very upset to see all that waste of

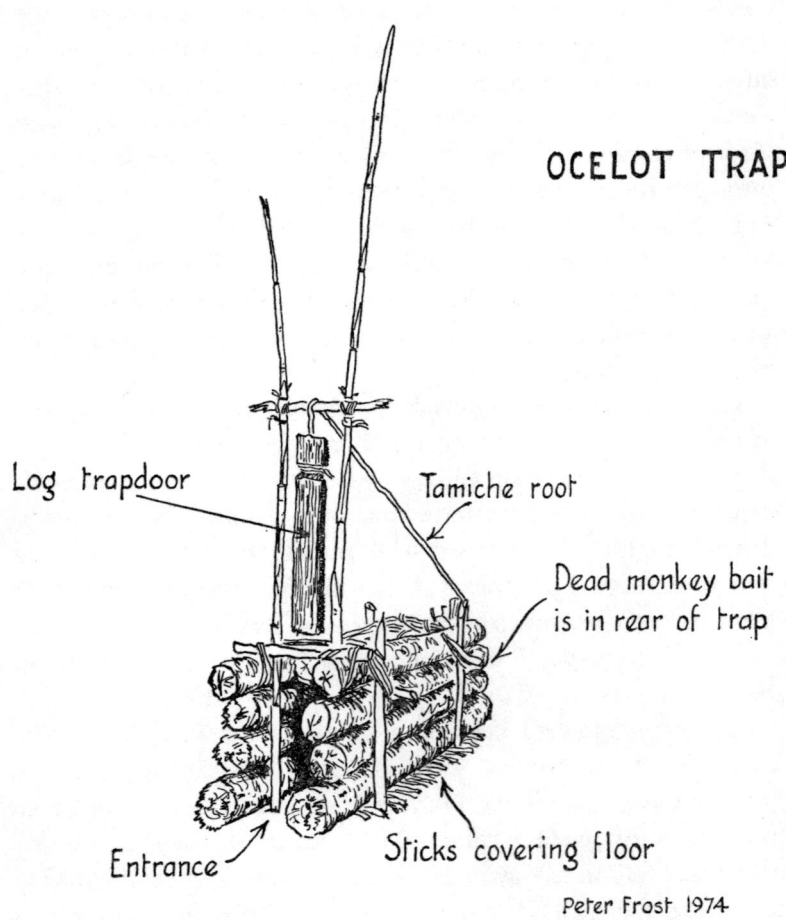

OCELOT TRAP

Log trapdoor

Tamiche root

Dead monkey bait is in rear of trap

Entrance

Sticks covering floor

Peter Frost 1974

good meat, not to mention the time and effort we'd spent on it. And within two days we had almost no food left except half a sack of rice, which would last out for a couple more weeks at best.

With the cold, grey weather we were having the outlook for hunting was bad, too. Still, there was no danger of actually starving; if the worst came to the worst we could always make it downstream on boiled rice. Meanwhile we decided to see if we could carry on as before, but now living almost entirely off the land. The whole area was swarming with a kind of bee—fortunately it didn't sting—it would settle on our arms and faces as we worked, sucking, apparently at our salty perspiration. This gave me the idea that we might find wild honey somewhere. When I suggested it to Wilfredo he said there was a bees' nest high up in a tree across the river, but he wasn't sure how we could get at it.

"I suppose we could cut the tree down . . . " he said hesitantly. But we never did become that desperate.

The day after Wilfredo's trap-building session Mario and I went off to pan up in the creek. Wilfredo decided to put in some time fishing. Without his boss chasing him Mario walked at a pace which was much more my style. He pointed out some inch and a half long *insula* ants. He said they had a fiendish sting which would put you out of action for twenty-four hours. Mindful of my perilous assignment to collect biting insects for the Royal Army Medical College, I dug out my specimen-collecting tubes and proceeded gingerly, with Mario's help, to shoo a couple of reluctant ants down the neck of a tube. Later, back at the camp, they would be pickled in alcohol and labelled to join the rest of our collection. During the journey whenever I was sitting with nothing to do I would try to catch insects that were biting me or threatening to, or ones that looked unusual. Ultimately I brought back 155 specimens and more than thirty different species.

As for our gold panning, it was a bad day. Neither of us had much luck. We came to the conclusion that there was no more gold for us there, so we packed up early and carried our pans and shovels back to the camp. Wilfredo had shot three Makisapa monkeys for us to eat, though he hadn't had any luck fishing. They prepared the monkeys by burning off the fur, and put two on a special frame we had made to smoke over the fire. The other went into the pot to boil. It would have to boil all night to be tender enough for us to eat for breakfast. And when the time came for us to eat that stew—my first taste of monkey—it turned out to be a little tough and stringy, but not too bad a flavour, a bit like mutton. Just as well, because we had two more to eat in the next few days.

That morning began with a tour of inspection for the ocelot traps. But they were all empty and still baited. The weather was a bit cold for *tigrillos* Wilfredo said. Then we went panning, this time not far from the camp, by the edge of the Rio Kipoznue itself. Wilfredo's tryouts of two days earlier had produced unexpectedly good results in this spot. And this morning's work was really good, with chips coming in almost every pan. When things were going so well one never moaned about one's backache or stopped to stretch. For the first time in days Wilfredo was singing away gaily, and he and Mario together amassed about six grams, while I collected about two grams. We still weren't going to make our fortunes at this rate—our combined take for the morning was worth about £10—but it was a pleasant way to spend the summer! At lunchtime Wilfredo was in excellent spirits and decided to tell Mario and I a sort of "Just-So" story he knew, as we ate our cold rice and chewed smoked monkey. He called it, *Why the Makisapa Monkey has only four fingers*. It went something like this:

Makisapa was swinging from tree to tree in the jungle, when he met up with Coto, a monkey relative.

"Hello, *compadre*," said Coto slyly, offering his hand. "How are you today?"

And he shook hands with Makisapa, hiding his thumb in the palm of his hand.

"Hello to you, *compadre*," said Makisapa. "But what's wrong with your hands? You've only got four fingers, Coto."

"Why, haven't you heard?" said Coto. "There's a new law out about monkeys to stop us from stealing. From now on we are forbidden to have more than four fingers. If you don't cut off a finger they'll catch you and hang you. Look here, I'm a good friend. Here's my machete, I'll do it for you."

Bang! Swash! Zang! Wilfredo's sound effects.

Off came Makisapa's thumbs. "Gracias, *compadre*, for saving my life," said Makisapa, and he went painfully on his way. And that is why the Makisapa has only four fingers when every other monkey has five.

Now we all know [continued Wilfredo, while Mario and I sat attentively; resting on the beach with our pans lying at our feet, and the sun beating down on us] that Makisapas and Cotos are enemies and have terrible fights whenever they meet. Well this is what happened: some weeks later Makisapa met Coto up in the trees, and Coto, forgetting himself, waved cheerfully at the other monkey.

"Coto," said Makisapa suspiciously, "how is it you've got five fingers again?"

"Oh, er, mine have grown again," said Coto. But then Makisapa realised he had been made an awful fool of, and—Oucha! Zwang! Zach! [more Wilfredo sound effects]—they had a terrible fight. And to this very day the Coto and the Makisapa—who now only has four fingers—will always have a terrible fight when they meet."

Wilfredo stood up. "Time to work again," he said, and started off digging away at the clay surface we'd been working all

15

morning. The afternoon didn't produce such good results. Gold really was a temperamental thing, disappearing at the merest whim. We never knew where we were with it.

At midday a Faucett DC–3 plane flew overhead towards Puerto Maldonado and returned an hour or so later. On my return journey I was to experience this same DC–3 journey. It is an unpressurised aeroplane – one of the world's great aircraft I'm told by people who know about these things – where the passengers sit in two rows on canvas seats, with their backs to the windows. When the aeroplane crosses the Andes you stick a tube in your mouth and breathe in oxygen. Among the passengers were two Indian women wearing the customary long and full Peruvian skirts, with plaited pigtails and high-crowned straw hats with black bands. These two ladies had a hen each, tucked under their arms, and when these birds got too upset with the altitude they would solemnly be given a whiff or two from the oxygen tube, to the enormous amusement of all the other passengers. But, at the same time there were, too, many passengers who had not really appreciated the value of the oxygen tube at high altitudes and they were paying dearly for it by airsickness to such an extent that there were no more brown paper bags to be had. There was no proper steward either. The crew consisted of two pilots and an engineer. The engineer was supposed to double up as steward, but either he was more urgently needed for his duties in the cockpit or he couldn't stand the mess. Anyway, he stayed prudently clear of the cabin throughout the journey.

We thought Katie, Oliver, Nick and Igor must be in that plane, just above us, getting ready to use the oxygen tube themselves on their way over the Andes. It was just over a week since they'd left.

In the morning as Wilfredo checked his *tigrillo* traps Mario and I set off to pan. Again I had a fantastic morning and in the afternoon, once again I didn't get a single chip. I just couldn't understand it because I was digging in the very same patch. Mario seemed to be lucky all the time, and had much steadier results than either Wilfredo or me. Wilfredo, though, when he ran into a lucky streak, invariably did extremely well. It was just like cards.

On Sunday we went fishing again and in the same pool where we'd caught our first monster, Wilfredo caught another enormous catfish. I myself got something really big on the line, which I could hardly hold. Trying to land it I braced myself against a rock and heaved, and the stupid line broke: we had lost one of our precious hooks. Wilfredo didn't quite know what to say. If it had been Mario breaking the line he would have raved, and called him *"Burro"* — donkey — and told him he was infantile and incompetent, that he had much to learn before he could consider himself a fisherman. As it was he just glared at me and went off to fish with the bow and arrow while I kept very quiet for the next hour or so.

In the morning I prepared to fry some fish for breakfast while Mario and Wilfredo went to inspect the *tigrillo* traps. The tapir fat had water in it and getting the fish to fry properly was impossible. When Wilfredo came back he took one disgusted look and said bluntly, "That's not fried fish, *señorita,* it should be crisp on the outside" (as if I didn't know!), but he was in a good humour because he'd got a *tigrillo.* They spent the morning skinning the animal very, very carefully, then they stretched it out flat using a frame made from strips of bamboo to hold it taut. Afterwards they set it out to dry in the sun for several days. Wilfredo paused to stroke the soft fur thoughtfully for a moment, before stepping back to admire the skin. I fancied one of the ocelot's gleaming white fangs to hang on a chain round my neck. Wilfredo said it was difficult to get out the tooth while the flesh was still on the head, and he wedged the skinned head between some rocks in the shallows for the little fish to strip. I didn't receive my fang, because next day we discovered that some great catfish or other had waltzed off with the entire skull! After that I was especially careful when washing my hair in the river.

There were many times when I marvelled at the depth of Wilfredo's knowledge and understanding of wild animals. He knew every animal call and bird song in the forest, could identify every track that appeared on the beach. And at times he seemed to like wild creatures better than humans. Whenever he told animal stories he gave the impression that he was talking about friends, people he knew well. He was an odd mixture of animal lover and hunter, and though he killed animals he never did so just for sport. An alligator took up residence near the camp for a while, and used to float for hours just under the surface, looking like a log, a little way downstream. I think any other hunter on the river would have shot it. But not Wilfredo; it wasn't doing us any harm, and he was content to live and let live.

In the afternoon Wilfredo and Mario started working thigh-deep in water near the camp. After a while they got really excited. They had hit an especially good patch, but I wasn't accustomed to shovelling out gravel from that depth of water, and I managed to spill most of it before I could get it into the pan. I left them to it.

During the night I woke sweating, feeling very hot and a little shaky. Later, though, I dropped off to sleep again, and in the morning I felt all right: I dismissed it as nothing. Any worries I may have had were pushed to the back of my mind by one of those funny-looking-back-on-it incidents, which happened immediately after I got out of bed. My spare trousers had been hanging up to dry, and while I slept they had been occupied by a colony of biting ants. I only discovered this when I put them on. Fortunately Wilfredo and Mario had gone off to inspect the traps, so they were too far away to hear my shrieks or witness my discomfort as I hurriedly undressed and flung the trousers on the ground.

When the boys came back we all went down to the river to pan. That was when I realised that there was really something wrong with me. I tried to dig and just couldn't fill the pan up I was so weak. I made it back to the camp to lie down, and slipped into my sheet sleeping bag, lacking the energy even to hang up my mosquito net. There I lay for most of the day, sweating, shivering, tossing and turning, sleeping fitfully, and generally feeling very rough indeed. Towards evening I developed an unquenchable thirst and began climbing shakily out of bed every few minutes to make myself yet another cup of tea.

Next morning I felt really weak. Wilfredo said I must eat to keep up my strength and thrust a huge plate of rice and beans at me, but I just couldn't look at it. He was getting worried about me by now and kept urging me to take some medicine. I tried to tell him that it was no use taking a remedy until I'd diagnosed the illness, but in the end I took a Panadol tablet just to keep him happy.

The boys went on panning that day, but at lunchtime I noticed Wilfredo and Mario both looked rather gloomy and uncertain. Wilfredo didn't know quite what to do with me. But after lunch he came up and said it was decided: we were going back. Even if my fever got better, he said, I would be too weak to go further upstream. As far as he was concerned the gold production here wasn't up to much and the *tigrillos* seemed to be avoiding his traps, so he was happy to get back and see how things were going at his farm.

I just lay back weakly and tried to smile and said:

"*Si*, Wilfredo." Two days earlier I couldn't have conceived turning back; I was anxious to press on and ready for several more weeks in the area. But the illness had quickly drained away all my strength and my enthusiasm. I was glad the decision had been made. Disappointment would come later.

I suggested we might turn upstream when we hit the Rio Colorado and try to pole as far as an Indian village I'd heard about. Some American missionaries had once put in a landing strip, and might well still be there, with their radios, and I would be able to get out quickly for medical treatment. I was beginning to feel as though that might be necessary. But Wilfredo vetoed the idea, saying that the chances were the radio would be out of action or the strip would be inundated with mud, in which case we would have lost two or three days. No, we would set off next morning for Camp Colorado.

So I spent the rest of the day slowly packing my things, in between longish periods of lying flat out, exhausted, on my sleeping bag. I was still drinking endless cups of tea, though I hadn't eaten anything for more than twenty-four hours. Every item of clothing I possessed was now soaked in perspiration, and most of them had already been in need of a wash when my fever started the day before, so I was feeling uncomfortable as well as sick. I couldn't wash anything because Wilfredo forbade me to touch water or get any part of me wet; he said it made the fever worse. I thought this was an old wives' tale, but I didn't have the energy to defy him.

Next morning Wilfredo did a final round of the traps, shutting each one so that no unlucky animal should wander in there to be caught where it would starve to death. Then he and Mario swiftly bundled everything up and loaded it into the canoe and we were off, me clutching a kettle full of tea to drink on the way downstream.

The river was very low and the boys had to work really hard. Every spot where the river was shallow or there were obstructions was even trickier than usual, especially with my extra weight to manhandle. I was in no condition to stand, let alone get out and push.

The next couple of days were a dim, hazy experience. Often the world would swim before my eyes, and I was glad to be lying down. My limbs ached. Sometimes I would sleep uneasily for a while, at others I would sit up taking everything in, feeling weak but not wanting to miss anything.

That evening we camped close to the river. In the night it

began to rain, and by morning it was cold. It was raining so hard
that no one stirred until 7 a.m. Then:

"*Creciente! Creciente!*" from Mario. The river was rising fast:
we *had* to get up. We threw everything into the canoe as fast as
we could, and by the time we pushed off the river had already
covered the spot where we'd slept.

"Well that's lucky," said Wilfredo. "We'll float down the
rapids now instead of pushing. It won't take half the time to get
back."

But I had a problem. I was freezing. I was swathed in plastic
sheeting to keep off the driving rain but inside I was shivering
like a jelly and my teeth were chattering noisily. Wilfredo said
he would stop as soon as the rain eased up and make me a nice
hot cup of tea. When he did stop to build a fire our brew-up
turned into a race against the river. The level was rising so fast
that it was a toss-up whether the kettle would boil before the
river swamped Wilfredo's fire. We won—just.

The tea warmed me up a bit, but I was soon feeling unbearably
cold again. Inspiration rescued me. I suddenly remembered my
"survival bag", something we'd bought from Pindisports for
emergencies. This was a kind of sack made of aluminium foil.
If you climbed in it'd retain all your body heat within the bag.
After about half an hour in my sack I felt quite warm and comfortable.

Later that morning we left the Kipoznue and rejoined the Rio
Colorado. As we did so Wilfredo sang a plaintive lament, making
it up as he went along.

"Kipoznue, Kipoznue, why were you so hard on us?
Why did you not give us more gold, more *tigrillos*?
Didn't you want us to stay with you?"

We hadn't seen anyone for two weeks, and twenty-three
days had passed since we'd waved goodbye to Wilfredo's men.

Here at the mouth of the Kipoznue we made contact with people once again. A group of Indians from the American mission was camped on the opposite bank. They let us heat a kettle on their fire, and then presented us with a short, thick length of bamboo stuffed with fish, in exchange for some rice. Wilfredo said that to cook it one only had to push it into the embers of a fire and leave it to bake.

Wilfredo was very cheerful now, singing and whistling, and telling us how much he was looking forward to seeing his dogs and cats and his chickens, and even his men. Further down the Colorado we pulled in at the camp of a friend of his, Isaac Margol. We'd stopped here on the way upstream, but we hadn't seen Isaac, only his wife. They gave us some hot chocolate to drink. Tinned milk! What a luxury. We accepted their invitation to camp here for the night.

More than three days had passed since last I had eaten, and I felt very much like a straggly weed. But I was over the worst of my fever, and was able to eat a few mouthfuls of the beans that were offered to us in the evening. That night I slept well.

Wilfredo took me aside next morning.

"*Señorita*, my friends have a little baby that is very sick. They don't think it's going to live. They were wondering if you could baptise it."

I stared at him. "I can't do that, Wilfredo."

"Please. It means a lot to them that the baby shouldn't die without being baptised. There are no priests here. Anyone in authority will do, anyone they look up to."

I was dumbfounded. Baptise a child? I hadn't the vaguest idea what to do. But Wilfredo was insistent. It would mean a lot to them, he repeated.

Isaac and his wife brought over the sickly child. It had been born just a month ago, and the chances were it would die here, unnamed and unknown, without appearing in any official register. I told them all the reasons why I shouldn't be doing this, but they wanted me to go ahead with the christening. Isaac poured some salt into a cup of water and his wife fetched a wild flower from the jungle. They prompted me with the words of the ceremony in Spanish *"in el nombre del Padre, Hijo ...* this child is named Jacqueline Margol" and everyone made the sign of the cross on the baby's forehead, first dipping the bright jungle flower into the salty water before the sign. They thanked me and Wilfredo said it meant a lot to them. They'd chosen the name Jacqueline after President Kennedy's wife, they'd told me. Since then I've often wondered if the little mite died, or if perhaps there's a small girl named Jacqueline Margol just learning to crawl along the banks of the Rio Colorado.

There was one more stop before we reached Wilfredo's place—the shop at the mouth of the Rio Pukive. There I bought several bottles of wine for Wilfredo, a long-outstanding payment on wagers of a bottle of wine per fall into the river on the way up the Kipoznue.

Back at Wilfredo's place there were all kinds of messages waiting for me—as soon as I arrived I was to go to Camp Colorado where they would arrange to take me downstream.

So it was all over. This time we hadn't made our fortunes. We'd found enough gold to pay off our workers and enough gold to gain respect from the local Bank. They hoped they would see us again soon.

I'd been away from Lima for ten weeks, and during that time I'd grown accustomed to this place, the river, the birds, the heat, and the atmosphere. And as for the gold, well that was in my blood by now. Once you've found gold, as all old prospectors know, you never give up. You *know* that higher up the river, deeper into the mountains, the Big One is there and that you can be the one to find it.

Some day, I knew, I would be back.

Appendix: *Equipment*

WE HAD SENT A large part of our equipment back to Camp Colorado, and from here on we would be less comfortable. But the equipment we had brought with us all the way from England had proved useful—often indispensable—and, for the most part, reliable. My brother was to write later, when I was back in England: "Your equipment and organisation was better than any other expeditions I've seen." I am quite pleased about this because he isn't in the habit of handing out gratuitous compliments to his sister; and he has taken part in several expeditions—and spent some time in the jungle himself, and reported on many others. So he knew what he was talking about.

We did have plenty of very useful pieces of good equipment, and we had a good general coverage of our needs—there were no enormous omissions. I can think of a few extra things we could have done with, but mostly luxuries like more shampoo and foodstuffs. As for organisation, we had taken some trouble to plan things out carefully and on the whole they went very smoothly, though part of that was due to luck rather than good management.

Four months before we were due to leave we had our initial basic support from the *Daily Mail*, and from our publishers.

This provided essential capital and without that the gold hunt would never have got off the ground. Once we had that, though, we knew we were set to go whatever happened, and all we had to do then was lighten the load on our budget as much as possible by canvassing many companies for extra support, in the form of free supplies, services and equipment. If I had sat down at the outset and written to everyone I could think of immediately, I believe I would have done much better than I actually did. As it was I only wrote sporadically, over a longish period, whenever I got a new idea for something we would be needing. Not until two weeks before we left, for instance, did it occur to me that we would need more than one stick of insect repellent and one tin of spray per head. But here I was lucky. Someone pointed out to me that Shell manufacture a range of insect repellents and I contacted them at once. They were most helpful, and sent round a good-sized parcel of sticks, aerosol cans and Shelltox, a coil which burns slowly when lit, producing a pungent smoke that drives away mosquitoes most effectively.

I tried to get us free air passages, but on this one we were out of luck. The airlines were all most pleasant, giving perfectly plausible reasons why they couldn't give us free or reduced passages, or even carte blanche to get our baggage through without paying overweight. It was all a question of their advertising budgets being low at that time of year, they said. But I still found this hard to understand, since many of the planes that fly to South America travel half empty. Initially I had been far too optimistic and rather too correctly I only contacted each airline after being refused by the one before, thinking it would be too embarrassing to have two say yes at once; One airline took two months to think about it before they turned us down. It was another case where I should have written to all the possibles at the start.

At the same time as I approached the airlines I did a lot of research into charter and cut-price air fares. I found that we could save a maximum of twenty per cent on the long haul to South America, by putting up with a little inconvenience of going a long way round on a slower plane. But we needed to save more than that. I worked out that the cheapest way for us to get to Lima in relative comfort—and not taking too long about it—was by booking a charter flight to New York and then renting a self-drive car to travel down the East Coast to Miami. I discovered one airline that offered a 150-day excursion return fare Miami–Lima at much less than the standard rate. We could stay our two or three months and still remain inside the excursion fare limit. But I did find that airlines seldom volunteered information on their cut-rate travel schemes; one had to dig to find out.

Although I championed hammocks as the perfect lightweight portable bed, I ended up wishing, along with Katie and Oliver, that we had brought camp beds. It was impossible to find out beforehand what kind of arrangement would be best for the area we were going to, and I know that in other areas not so far away we would have been miserable indeed without hammocks—travelling on some river boats for instance. It was just bad luck, and the only way to have guaranteed our comfort would have been to take both.

We took with us an expedition tent which we'd bought at Pindisports. It weighed a mere five pounds, and for that extra weight it was worth taking anywhere just as security for any eventuality. According to the publicity blurb it would sleep four people. On some glacier in the high Andes it would, but in that sticky tropical heat it was unthinkable to try sleeping even the three of us in there. We each had a survival bag too. A small packet four-inches square containing a light aluminium bag which when unfolded was the size of a sleeping bag.

Nick lent us an inflatable tent he had for his family, and that proved to be excellent for storing all our food and equipment during the early part of the journey. We also used it as a changing room and for washing when the weather was cold and we didn't fancy a dip in the river.

Hutchinson's in Paris air-freighted us a rubber dinghy, with a set of oars and an air pump. One of our biggest disappointments of the whole journey was that this gift sat, trapped in Peruvian customs because of a delay in receiving the correct papers from France and we were never able to use it.

Apart from this one problem we had no trouble with the customs and in this we were exceptionally lucky. A British mountaineering expedition which arrived at more or less the same time as us was delayed for over three weeks before they could set off from Lima; and they had to leave behind a huge deposit to make sure that they took the equipment home with them instead of selling it in Peru.

I wrote to Timex to see if we could all have free watches. We all had watches of our own, but mine certainly wasn't waterproof. They wrote back very promptly and cheerfully saying they would be delighted to help us. These watches were easily the most popular piece of equipment we had. They had impressive looking faces with plenty of different dials and almost every person we met on the river wanted to buy them from us. Oliver presented his to Villanueva, and I was to give mine to Wilfredo as we parted. Katie was the only one to arrive home with hers and she gave it to her fiancé.

I asked Mary Quant Cosmetics for their advice on face creams and they offered to supply us with several products which Katie and I used daily in a fight against tell-tale wrinkles!

Our visit to Pindisports in Holborn had taken one whole afternoon. The shop is all set out in a semi self-service fashion so it was easy to find everything we wanted. The only items we took that we hardly used were our compasses; and this was mainly because through carelessness and lack of foresight we kept forgetting about them. More than once when we had lost our way in the forest they could have saved us a lot of time and anxiety. We bought whistles to signal in case we got lost, but they were only ever used for refereeing the local football matches. Nick had said he thought they were a good idea, but locally they were rather laughed at. People pointed out that we could shout just as loud as we could whistle. Our waterproof torches were invaluable and treasured possessions. Another time, though, I will check the beam of each torch as I buy it, to make sure it casts an even, consistent light. Our torches were all the same make, but one of them threw a beam with a small dark dot in the middle like the centre of a bullseye. It was no use for spotting alligator's eyes at night.

We chose a set of pots and lids that fitted inside each other like a nest of Chinese boxes. These were very practical and we used them all the time, until we started cooking for more than just the three of us. The lids fitted so tightly that we could store leftovers in them without worrying about insects crawling in. But I completely forgot to get utensils for stirring or ladling, so that we usually either burnt the pans, or burnt our hands trying to stir with short spoons. We took both plastic and enamel mugs with us. Katie and Oliver preferred the enamel ones because they thought the plastic ones gave the drink a synthetic taste, while I used the plastic mugs because I like drinks very hot, and the enamel mugs used to burn my lips.

The inner sleeping bags my sister-in-law Consuelo made for us turned out to be more than just a luxury. Mine at least was indispensable, as when we were on the Rio Kipoznue I had a bad fever and this was easily the best thing for me to wrap myself in, as it didn't take too long to dry out. Katie and I decided to get sleeping bags without a zip. Personally I don't like the scratchy feeling of a zip and if it gets cold a zipped bag is draughty. If you have a zip you can open out the sleeping bag to dry more quickly, but then the sun was so hot we had no problems on that score.

Our life-jackets and our camera bags we bought from the

Canoe Centre. The jackets were called buoyancy aids, which are different from the inflatable type they have in aircraft, which need air inside them. We were advised by a sports photographer friend that Harishok were the best kind and most canoers used them. They were splendid, and we should have used them more than we did. It was a mistake to send ours downstream (though one was kept back for Igor) when we split up our equipment before setting off up the Rio Kipoznue. Our camera bags were inflatable, and therefore — we were assured — would float when inflated even with the cameras inside. Happily we never had to test this assumption. At any rate, we had the cosy feeling that if the canoe went over the cameras would be dry and safe. They were particularly good, too, in the canoes where there was always water slopping about — especially as it was quite easy to get at the camera quickly. We kept our cameras in these bags all the time, as a precaution against the humidity that could so easily have ruined this equipment. Due to rough treatment we punctured the bags once or twice but Oliver mended the holes with a repair kit.

We each had a machete we'd bought in Florida — Shefeeld Steel it said. It was only later we noticed the spelling and the tiny letters underneath, Made in Japan. We used these machetes regularly though they chipped and bent easily. I used both my sheath fish knife and my pen knife several times a day.

Hats are an essential garment for this type of journey, as I knew in advance, and we really needed two each in case we lost one. Oliver borrowed one of mine when his blew into the river, but it didn't have a wide enough brim and his nose went bright red and peeled as a result.

Katie made us each sewing kits before we left, but afterwards we decided we should have included in this, a spare zip each, extra buttons and much more cotton. We had no idea just how much sewing we would need to do. Some thick nylon thread, that Katie had included, for instance, was all used up on holding together rotting gym shoes. What with the permanent humidity and dampness and having to wash clothes so often, the seam stitching started to rot very quickly and there was something to mend every day.

I had been told by a friend that the screws on the Nikon camera easily worked loose so I should take a set of camera screw drivers with me. I did and frequently checked the cameras, tightening up the loose screws. I should have given my glasses the same treatment because the screw worked loose on the nose pad without my noticing and one fine morning it just wasn't there any more. Luckily Katie is a very handy person. She carved me a wooden replacement with her trusty knife and bound it on with thread. The repair lasted very well.

We had three sets of clothes each, but we really needed four. If the weather was bad we couldn't keep up with drying them out. We definitely needed three sets of footwear, too, which Katie did have, but Oliver's and my final pairs somehow were mislaid. Although we set out with a preference for thin trousers that would keep us cooler and dry out faster, the pairs we wore most were the thicker pairs that the mosquitoes couldn't bite through so easily. The same applied to our shirts and Katie and I even wore cotton T-shirts under our blouses to cut down the number of bites. We all had a belt and we couldn't have done without them as we got thinner with each week that passed.

Our medical supplies covered a wide range of ailments and a colleague of my mother's, Sev Solomon, a Brighton pharmacist, took a lot of time and trouble to get us what we needed and advised us on essentials. We had plenty of antibiotics in pill form, emergency injections against allergies, and many lesser preparations: eye drops, antihistamine creams, antibiotic creams, diarrhœa pills, constipation pills, remedies for toothache and earache and anything else we could possibly imagine we might need. A lot of these rugged chaps who go on expeditions equip themselves to cope with only dangerous illnesses. Personally, I think this is a rather masochistic attitude, and I didn't want anyone suffering pain needlessly, even if we weren't going to die from a headache.

We had vast quantities of footpowder and yet we used it all up. I wonder now what we would have done without it; I think our feet would have just rotted away altogether with the amount of soaking they suffered.

In spite of all our precautions, when we fell ill our medical knowledge was insufficient to deal with it. We thought Katie had food poisoning—but we weren't sure, and we didn't really know what to do about it. Neither did I know the cause of the fever I had. I stuck to what I thought was correct in starving a fever and feeding a cold but next time I'll find out more about curing fevers before I go.

H. J. Heinz have a section of their company called Heinz Erin which makes dried foods and meals in a light package, and they offered to supply us with some items which were quite the most delicious packet foods I have ever tasted. The dishes like chicken curry and beef risotto were so good it looked as though we'd spent hours preparing the meal. The desserts, too, were very good and practical at the same time, because they were so easy to make. Oliver was always happy to have the butterscotch flavour, but alas, they were all gone far too soon. The Creme Caramel that we ate on Katie's birthday was a lone survivor that had been hoarded specially for some time.

From Batchelors we bought various kinds of dried vegetables, dried minced meat, dried onions—very important for adding flavour to our food—soup mixes, macaroni with cheese sauce and dried apple. For ordering they sent an excellent chart with details of calories per ounce per product. But I didn't count calories and decided I'd have to settle for the rule of thumb, which was to order enough meat, vegetables, and vitamins to sustain us and keep us from being hungry. Even then, when boxes full of these foods arrived, we wondered how we would ever get them all to Peru, and gave some packets away. There were times later on when we felt it would have been worth spending any amount of money on overweight to have had those extra packets with us.

I like food, and if I seem to put too much emphasis on it I can only say that for us it was usually worth the trouble. It always made a bad day seem a little better if there was a good meal waiting at the end of it.

Our lightweight anoraks were especially useful. It might have been worth buying the trousers that go with them but then on the Amazon expedition two years previously, I took a pair doggedly from start to finish without ever needing to unfold them.

We should have taken with us some games — cards, back gammon, draughts, and so on, because there were days when it rained and there was nothing else to do. At times we were able to borrow them from other people, but it wasn't the same as having our own.

My attitude throughout the preparation of the journey was that we should primarily aim to enjoy ourselves. As we didn't plan to carry our equipment for long distances, I saw no reason for us to cut down drastically on items that would normally on such a journey be termed unnecessary, but which would be pleasant to have around. So if any of us particularly wanted to take something we took it. If we had tried hard we could have cut down on plenty of items, but we weren't trying to beat any records for hardiness and endurance.

There's a land where the mountains are nameless,
  And the rivers all run God knows where;
There are lives that are erring and aimless,
  And deaths that just hang by a hair;
There are hardships that nobody reckons;
  There are valleys unpeopled and still;
There's a land—oh, it beckons and beckons,
  And I want to go back—and I will.

from *The Spell of the Yukon*
by Robert Service

Epilogue

A FEW WEEKS AFTER I left the Kipoznue paradise an influx of petroleum prospectors arrived in the area, based on Manu but taking in the areas of Rios Upper Madre de Dios, Colorado and Kipoznue. They now have over five hundred people in the area and for instance they have eighty hunters working to find enough food for the group. They have helicopters and small planes to aid them in their search.

I should be happy that for several weeks I had the opportunity to live in an area that was entirely unspoilt by Progress. I hope they don't find oil there or anywhere near, but from the way they were talking when I met them I fear that the Kipoznue will not be the same when I return.

Index

Index

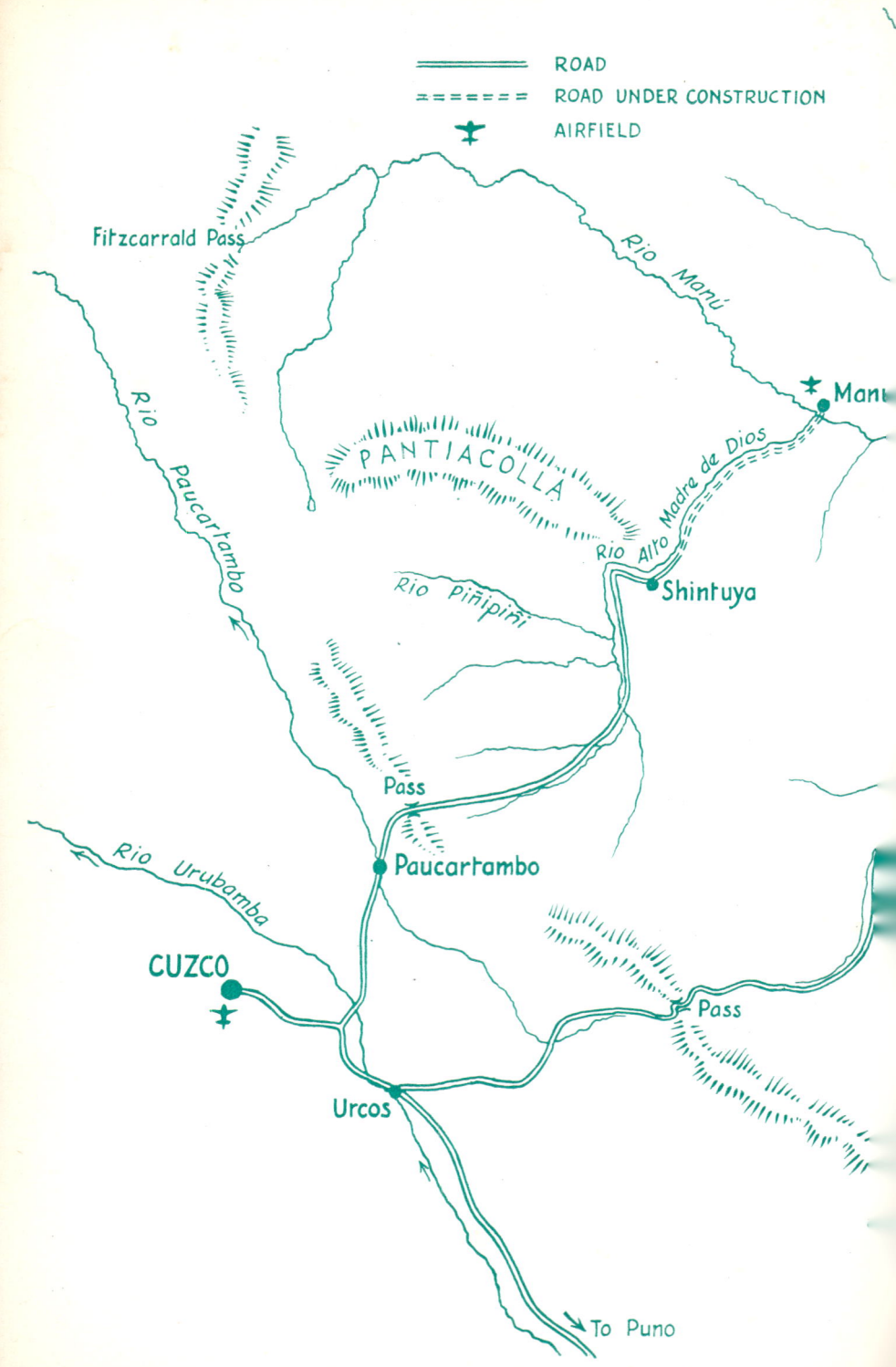